IDRC-082e

Computer Simulation of Soil-Water Dynamics: A Compendium of Recent Work

Daniel Hillel

This work was carried out with the aid of a grant from the International Development Research Centre, Ottawa, Canada.

Postal Address: Box 8500, Ottawa, Canada K1G 3H9
Head Office: 60 Queen Street, Ottawa

Hillel, D.
IDRC
Computer simulation of soil-water dynamics; a
compendium of recent work. Ottawa, IDRC, 1977.
214p. diagrams, graphs, tables.

/IDRC supp CRDI/. Monograph on the application
of /computer/ /simulation/ /model/s to the
/systems analysis/ of /hydrology/cal relations -
discusses the construction of /mathematical
model/s, and directives for /computer programme/ing;
presents models of the /evaporation/ process, /soil/
/water/ storage and conservation, hydrology of a
sloping field, and soil-water uptake by /plant/ roots.
/Bibliography/.

UDC: 626.8:631.4 ISBN: 0-88936-119-3

Microfiche Edition $1

Dedicated to the memory of my mother Sarah
who bore tragedy and adversity
with grace and undiminished faith
while retaining, and transmitting,
the joy of life and the love of learning.

ACKNOWLEDGEMENTS

Dr. David Hopper, President of the International Development Research Centre, and Mrs. Ruth Zagorin, Director of the Centre's Division of Social Sciences & Human Resources, for their encouragement and support.

Dr. Jan Feyen of the Catholic University of Louvain, Belgium, for hosting me most kindly during the early stages of the manuscript's preparation.

Dr. George Hornberger and Mr. Roger Clapp of the University of Virginia, Department of Environmental Sciences, for their helpful criticisms.

Mr. Jeffrey Kloss, for photographing the drawings.

Mrs. Janie Cohen, for her meticulous work in typing and improving the manuscript.

Last but not least, my family — Rachel, Adi, Ron, Sari, Ori, and Shira — for suffering the effort without complaint, if not entirely in silence.

TABLE OF CONTENTS

PROLOGUE

The needs of growing populations and of developing national economies have in recent years brought about great intensification of land and water use in a continuing effort to increase agricultural production. These needs are most pronounced in arid and semi-arid regions, where pressure on limited soil and water resources is felt most acutely. Numerous technological innovations have been applied in the field in an attempt to enhance the efficiency of soil and water management. The urgency of this task, however, has at times led engineers and developers into the pitfall of hastily adopting inappropriate methods of irrigation, fertilization, tillage and pest control which may eventually do more harm than good. Injudicious management of soil and water can cause waste of water and energy, and deterioration of the soil through salinization and erosion. Moreover, careless management in the agricultural field can result in serious environmental consequences of which we have only lately become aware. It is the role of agricultural and environmental research to generate basic knowledge of the complex system to be managed and to apply such knowledge toward the optimization of all controllable factors so as to achieve a higher level of production on a sustainable basis without damaging the environment. Scientific research, fundamental in conception yet practical in outlook, is especially needed to adapt or devise methods of soil management for the specific needs of rain-fed and irrigated agriculture in semi-arid regions, where even slight modification of the water economy and energy relations can sometimes spell the difference between sub-marginal subsistence farming and profitable production, and where ground water and surface-water pollution is an ever-present hazard.

Some of the fundamental problems of agricultural and environmental research are how to obtain knowledge of specific processes within a complex system of interacting and interdependent phenomena, and then how to reintegrate such knowledge so as to obtain a comprehensive understanding of the way the system as a whole operates. Such understanding is essential if we are to generalize experience gained under specific conditions and extrapolate it to different locations and seasons.

In recent years, mathematical modeling and simulation techniques, relying on the use of high-speed computers, have been developed for the purpose of providing a comprehensive quantitative description of the behaviour of dynamic systems. (Herein, we use the term *simulation* in the narrow sense to denote the construction and operation of numerical models for the purpose of testing theories regarding the behaviour of the natural system.) When successful, computer simulation can help us to reduce a seemingly incomprehensible system of almost hopeless complexity to manageable, orderly proportions. In the process of designing, operating, and attempting to validate a simulation model, we gain insight into the workings of the complex natural system and develop criteria for predicting its future behaviour under varying conditions.

Simulation techniques are applicable to the analysis of water, solute and energy transport in the soil, and to the utilization of soil moisture by plants under various climatic conditions. Mechanistic simulation models of this system are based on the mathematical formulation of mechanisms, such as those of the physical and physico-chemical processes, known to occur in the soil at rates which depend on the state of the soil system as a whole and of its component and interconnecting parts. Calculations obtained from the use of such models can allow us to predict how certain controllable factors (*e.g.*, irrigation frequency and intensity, water quality, soil surface conditions, etc.) can affect the pattern of soil water storage and utilization by crops.

The aim of this monograph is to describe the formulation of a number of models simulating soil physical processes and to illustrate the results obtainable from them. Although the presentation is based largely upon recent work in which the author was directly involved, he wishes to cite, as antecedents to this work, the pioneering contributions of the Dutch group under the leadership of C.T. de Wit; as well as of other numerous scientists, many of whom are included in the list of references.

The book is addressed mainly to problem-oriented research workers who are concerned with agricultural and environmental aspects of the soil system and who are interested in a theoretically based rather than completely empirical approach to research. A basic working knowledge of calculus and some, at least minimal, familiarity with computer programming are assumed. A few of the philosophical, as well as practical, aspects of simulation are considered.

We believe that the most crucial decision in any research project is made at the outset, and that is the decision what to research. We all suffer from limited resources of funding, manpower, and time, and when we decide to commit these to a particular experiment, designed in a particular way, we quite inevitably give up the chance to pursue other, perhaps more promising, directions. Hence we ought to pre-search for some predictive leads ahead of time, as well as prepare a way for evaluating the information we expect to obtain in the end. The author hopes to spur the interest of students of soil-plant-water relations in the possibility of applying simulation techniques as an extra dimension in their investigations. He does not wish to convince the reader of the lasting validity of the models presented herein so much as to challenge him to do better on his own. In science, and particularly in modeling, no "last word" (even if it is, for a fleeting moment, the "latest word") can ever become the "final word."

"Nature is a labyrinth in which
the very haste you move with
will make you lose your way."
Francis Bacon

"Although this may seem a
paradox, all exact science is
dominated by the idea of approximation."
Bertrand Russell

"Man was not born to solve the
problems of the Universe, but to
put his finger on the problem
and then to keep within the
limits of the comprehensible."
J.W. von Goethe

INTRODUCTION
FUNDAMENTAL PRINCIPLES OF MODELING AND SIMULATION

A. The Concept of a System

This book is concerned with developing an understanding of the dynamic physical behaviour of the natural system known as the soil. We have already used the word "system" as if its meaning in this context were self-evident. It is not. The word "system" is used so often and so loosely to designate a variety of concepts that it is hard to assign it a specific definition. As used herein, a system is a part of the universe which can be distinguished from its surrounding environment by either physical or conceptual boundaries. An animal, for instance, is a distinct entity within recognizable physical boundaries, which interacts with its environment through the exchange of energy and matter. Often, the boundary we recognize is merely arbitrary (as, for instance, the boundary we may wish to set between the soil, which is the subject of our study, and the subsoil, however it may or may not differ from the soil), as in nature the entire universe is really an integrated, continuous system. However, if we are to separate out a portion of the universe and treat it as a system in itself, we must have some definite criterion for distinguishing what belongs in the system and what constitutes an external effect (*i.e.*, a stimulus or disturbance originating outside the system).

Another attribute of a system is that it is composed of interacting parts. In an animal we recognize organs with specific functions. In the soil-plant system, we distinguish between the soil and the plant, and even in the soil itself we can often discern distinct layers. A major part of a system is called a *subsystem,* and elemental parts of subsystems are called *components*. Of course, the hierarchy of components, subsystems and systems can never be absolute, since even the most elemental part of a system can in turn be analyzed in such detail that it would become a complex subsystem. While recognizing the parts and their functions, our ultimate concern is with the operation of the system as a whole.

For a long time the dominant tendency in our science had been to focus upon individual phenomena or processes which could be studied in isolation. This was, and in fact remains, a necessary endeavour in the development of our understanding of the system's mechanisms. However, exclusive concentration upon artificially isolated aspects of the system can lead to disregard of important other aspects (such as interactions and feed-back mechanisms) and thus to oversimplified conceptions about the way the system operates. Now our task is to assemble and integrate our fragmented knowledge of the system's component part in order to develop a more comprehensive conception of the system as a whole. We shall attempt to do this by means of *models*.

Systems analysis is a term applied to the logical organization of data and theories concerning the behaviour of various systems into models, and the rigorous testing of such models for the purpose of validating and improving them, and ultimately of using them to predict the future behaviour of the systems which they represent.

B. The Concept of a Model

A poet gazing through his window may perceive the green field stretching outside as a realm of supreme serenity and quietude. But to the environmental scientist, it is a system in a state of incessant flux, where matter and energy are transformed and transported to and fro in a series of numerous concurrent processes involving physical, chemical and biological changes. Our mind boggles at all the many processes and variable rates which we know to be occurring all the while, and there may be many more processes at any moment of which we are totally unaware.

The real world, or indeed any perceivable system within it, is altogether too complex for our limited intellect to comprehend or to define in its entirety. In dealing with any particular problem, therefore, we are obliged to take the easy way out, which is to imagine the system to be simpler than it really is, by considering only aspects of it which pertain to the problem at hand. In so doing, we deliberately ignore other aspects of the system which, as far as we are concerned at the moment, are only irrelevant complications.

Such a simplified, and hence more readily definable and more easily tractable, version of reality is a *model*. It can be physical, as in the case of scale models of aircraft used for testing aerodynamic designs; or abstract, as in the case of conceptual or mathematical models.

Perhaps the most basic exercise in modeling is the development of human language. If an event is observed by one individual and not by another, then the observer can convey, using words, a selective description of his observation to the listener, who can then re-create the event in his imagination and thus perceive approximately what is happening. The selective aspect of lingual representation becomes apparent if we imagine a caveman warning his tribesmen: "many strangers are approaching!" In so saying, he does not bother to add that the sky is blue and the grass is green, facts which he has also observed but probably considered unimportant for the moment. Other primitive exercises in modeling are the carving of idols representing forces of nature or voodoo dolls representing persons. Viewed in this sense, modeling is seen to be a primeval function of the human mind in its perception and communication of reality.

Scientific modeling, to be sure, is a considerable extension, or refinement, of primitive modeling, particularly in regard to its exact quantitative formulation of interrelated events. Yet all models and theories remain approximations; they do not correspond to the observable facts in any obvious one-to-one way. For instance, we speak of heat "flowing" as if it were a fluid, which it is not. But the analogy to a fluid helps us to grasp the behaviour of heat, to perceive it in familiar terms, and it facilitates our speculations. The visual image of something flowing leads us to think of gradients, of a natural tendency to flow from a higher to a lower level. This has turned out to be useful and to correspond with some important facts.

However, when a model begins to depart too grossly from the facts (as happens especially when it is carried beyond its limited context) it becomes misleading, and we must modify or replace it. The complications which we have disregarded in any particular model do not in fact disappear. Having once defined the seemingly most-important ("primary") effects, sooner or later we find that to refine and generalize our model we must now consider the next-to-the-most important ("secondary") effects, and so *ad infinitum*. Our developing knowledge of any complex system is achieved by successive approximations.

The principles we have thus far described in general terms apply specifically to the theories and equations employed in soil physics. At different times and for different purposes, soil physicists have compared the soil to a collection of spherical particles, or to a bundle of capillary tubes, or to a collection of parallel colloidal plates acting as electrostatic capacitors, or to a mechanical continuum with elastic or plastic properties; in other cases they have assumed the soil body to be spatially uniform and even one-dimensional. We must be careful not to take these theories too literally, as they were borrowed from simpler or "purer" systems by entirely fallible (though undoubtedly courageous) scientists attempting to make their inherently complex system manageable by simplifying it. As this science, among others, keeps developing, its tools are becoming more sophisticated and capable of accounting for complications which previous soil physicists perforce disregarded. A case in point is the use of high-speed computers to describe varied phenomena in heterogeneous soil media, as we hope to show in this monograph.

C. Mathematical Models

Scientific models are best expressed in the concise, terse, objective, universal and flexible language of mathematics. We set an equation which describes how the system behaves, in accordance with the best available evidence. Next, we transform the equation to anticipate how the system should behave under changed circumstances. In so doing, we not only summarize what we already know about the system, but also project into what is not yet known. That is to say, we predict quantitative relationships which we have not yet measured. We then check our predictions by experimentation. If the results fit, we have a working model. If not, we revise our model and try again. Thus, theory cannot advance without experimentation (or, at least, without systematic observation). Conversely, experimentation without theory is likely to be sterile and pointless, as it might hopelessly bog us down in an ever-deepening mire of seemingly unrelated and random facts.

Following are several definitions of types of models which are not necessarily mutually exclusive. A complex model may encompass several of these in conjunction.

Empirical models are based on observed quantitative relationships among variables without any insight into the functional or causal operation of the system. The formulation of an empirical model is perhaps the earliest and most primitive stage in the development of any science. Thus, some early soil physicist of the 19th century, working in location X, may have noticed that soil temperature is related to day length, whereas another worker in location Y may have observed that it is related most strongly to soil moisture, even while their contemporary in location Z postulated that it is related primarily to soil color. Each could then formulate an empirical equation to approximate his own observation, but none could develop a comprehensive and general model, incorporating all three effects, without knowledge of the processes governing the energy balance of the soil.

Stochastic models are those in which one or more of the functional relations depend on chance parameters, and are hence related to a probability distribution. Thus, any seasonal prediction of soil moisture under natural conditions would presumably have to relate to the probability functions governing the time of occurrence and quantity of rainfall.

Deterministic models are non-stochastic in the sense that no random variables are recognized. Exact relationships are postulated, and the output is predicted by the input with complete certainty. Thus, the content of a reservoir can be calculated with certainty for all times when the initial content, as well as the input and output rates, are known.

Mechanistic models are based on known mechanisms which operate within the model, such as the fundamental laws of physics and chemistry. As such, mechanistic models are often a subclass of deterministic models.

Analytical models are ones in which all functional relationships can be expressed in closed form and the parameters fixed, so that the equations can be solved by the classical methods of analytical mathematics.

Numerical models are such in which the governing equations are solved by means of step-by-step numerical calculations, generally necessitating the use of a computer.

Continuous models portray continuous processes, in contrast with *discrete models* which include discontinuous or abrupt phenomena.

Dynamic models portray time-dependent processes (as opposed to static, or time invariant systems). Time is, of course, a monotonically advancing, irreversible, independent variable.

D. The Scientific Basis of Computer Modeling

Since time immemorial, man has been driven by a desire to understand how nature operates and thereby to predict the future course of natural events. The speculative philosophers of Greece tried to achieve this coveted power of prediction by purely deductive methods. They sought scientific explanations by simple analogies with experiences of everyday life, and questions relating to the theory of knowledge were answered in terms of picture languages rather than by rigorous logical analysis (Reichenbach 1951).

Francis Bacon was apparently the first to recognize the limitations of speculative philosophy as a methodology for predicting the future. In his book *Novum Organum* (1620) he argued that reason by itself cannot have any predictive capacity, and that only the combination of reason with observation can create such a capacity. Moreover deductive logic must be augmented by inductive logic so that we may reason both from the general to the specific and vice versa.

Based on this principle, Bacon enunciated the so-called scientific method, which consists of four steps: 1. Observation of the real system in operation. 2. Formulation of an hypothesis (a mathematical model) to explain how the system operates. 3. Prediction of the system's behaviour on the basis of the hypothesis (by obtaining solutions to the mathematical model). 4. Performance of experiments to test the validity of the model's predictions.

These principles are very well known universally, and the reader may wonder why we have seen fit to repeat them here. We have done so to emphasize that computer modeling can do nothing more than follow the scientific method. It is not different in principle from what scientists have been doing for centuries; it merely introduces a powerful new tool into the process. As the microscope and telescope have served to extend the capacity of the human eye, so the computer can serve to extend the capability of the mind's ability to calculate rapidly and to account for simultaneous phenomena. It is not, we might add, an extension of the human mind's faculties of imagination or intuition or ingenuity, qualities which, for better or for worse, cannot be computerized.

E. Computer Simulation of State-Dependent Systems

Computer simulation can be defined as the construction of mathematical models to imitate the essential features and behaviour of a real system, the adaptation of such models to solution by means of a computer (analog, digital, or hybrid), and the study of the properties of such models in relation to those of the prototype system. Alternatively, and perhaps more specifically, we can define simulation in our present context as a numerical technique for conducting hypothetical experiments on mathematical models describing the quantitative behaviour of dynamic systems.

Dynamic simulation models of the type we shall present in this book generally possess several essential components, including the following:

Exogenous variables are input variables, independent of the internal state of the system. They represent external factors which are imposed upon the system and, acting on it, induce changes within it. Such factors are often called "forcing functions" or "driving functions." As such, they can be either uncontrollable or controllable. Thus, for example, the pattern of insolation (or solar radiation) reaching the field is an uncontrollable factor, whereas the pattern of water supply, in an irrigated field at least, is a controllable factor.

Endogenous variables are dependent, output variables ("responses") of the model, generated by the effect of the exogenous variables on the system's state variables. Examples are the pattern of evaporation or drainage from a soil profile.

State variables are those which characterize the state of the system and directly determine the processes which bring about changes in the endogenous variables. Examples of state variables are soil temperature or water content, the distributions of which in a soil profile determine the conduction of heat and water, respectively. State variables are, in a sense, the intermediaries through which exogenous variables eventually influence the endogenous variables.

Rate variables control the rates at which various responses are generated by various changes in the system's state. Rate variables are often called *parameters*. As such, they represent the coefficients of the governing equations or laws describing the functional dependences of the endogenous variables upon their controlling state variables.

If, for an overall example, radiation is an exogenous forcing function and surface temperature is a state variable, then soil heat flux is an endogenous response variable and soil thermal conductivity is a parameter. An additional response to a change in surface temperature is the emission of heat by the soil surface, for which the rate variable or parameter is the emissivity.

Whether a particular variable is classified as exogenous, state, or endogenous depends on the purpose of the model. Thus, if we wish to calculate the influence of changing air temperature on soil temperature under conditions in which the former can be considered independent of the latter, we take the former to be the exogenous one. On the other hand, we can conceive a situation in which soil temperature is the primary factor causing a change in air temperature (as, for example, in a glasshouse), in which case the endogenous and exogenous variables change roles.

Before we can begin to operate a simulation model, we must also specify the structure of the model, including its geometric features, dimensions, and internal sections. We must also specify the boundary conditions of the model. Finally, for dynamic models, we must state the initial conditions of the state variables. The dynamic system being simulated is assumed to be "running," and hence continuously changing. Our simulation program is assumed to take over control of this "running" system at a particular point in time. The initial values we specify are thus based on the state of the system at the moment we choose to join the train of events, either as non-intruding observers or as active, purposeful, and systematic manipulators of factors.

A major principle underlying the operation of a state-dependent system is that simultaneously occurring events do not affect or depend on each other directly. Each event is affected only by the controlling state variable at the beginning of each time step. Therefore, if we know the conditions of all state variables as each time step is begun, we can determine all state-dependent processes in parallel. The order in which these simultaneous processes are computed does not matter, as each refers to the same values of the state variables. During each time step, of course, each process can affect one or more state variables, but the latter values are not updated until the time step is completed for all the simultaneous processes. Only then is the overall or net effect summed up and each state variable is updated for the beginning of the next time step. The simulation process is driven by an internal clock which stops periodically to perform calculations, then advances in a discontinuous spurt only to pause again for the next set of calculations, etc.

We must remember that a digital computer cannot perform any abstract analytical operations with mathematical symbols as such; it can only reckon with numbers. A digital computer can add, subtract, multiply and divide; it can remember numbers and instructions, and it can follow a set procedure or program of calculations. Since it makes one computation at a time (though at enormous speed!) it is inherently incapable of operating continuously.[1] Since many dynamic systems can only be described in terms of differential equations, this is a basic flaw of digital computers. However, a digital computer can be made to *imitate* continuous processes such as integration, and various numerical techniques for this are now available as standard computer subprograms. The user himself may remain oblivious of the discon-

---

[1]Continuous integration can of course be performed by an *analog computer*. However, this advantage is offset by the difficulty of programming an analog computer, and by the fact that such systems are not as universally available and accessible as are digital computers. In principle, the best arrangement is to have an analog and a digital computer operate in tandem, so that the former can perform integrations while the latter performs arithmetic operations and makes logical decisions. Such a system is known as a *hybrid computer*. But at the present time, hybrid computers are still rare, expensive, and difficult to program.

tinuous nature of the computer's on-again, off-again operation, provided of course the time steps are short enough (yet not too short, for the user must eventually pay for the computer time!)

The method by which a computer is made to advance in time during a dynamic simulation can be based either on fixed time increments or on variable time increments. The latter method is likely to be more economical in terms of computer time when the processes considered vary considerably in rate, especially if they are relatively fast during a fraction of the time but slow most of the time. Under such circumstances, the computer can be made to adopt short time steps during periods of rapid change, and much longer time steps (a more economical procedure) during periods of slow change.

Once we have defined a system both geometrically and parametrically, formulated its mode of operation (generally by means of governing equations), and defined the values of its state variables at some initial time, we can predict the future values of the endogenous (response) variables if the future course of the externally imposed input variables (environmental factors) are also known.

The calculation procedure is straight-forward in the case of non-hysteretic processes, in which each time-step is determined only by the present state of the system. In some cases, however, the process is known to depend not only on the present values of the state variables, but on their past values as well (or, at least, on the direction of change which brought about their arrival at their present state, whether by decrease or increase of their previous values). Such processes are *hysteretic,* and will be illustrated in a succeeding section of this monograph, pertaining to evaporation under a daily cycle of evaporativity.

Figure 1 illustrates the concept of a state-dependent simulation model which can, in principle, be operated upon in at least three ways, or phases: (1) *Construction:* If we know, after observing and experimenting with the prototype real system, the corresponding past histories of the input variables I, of the state variables S, and of the output (response) variables R, we can try to establish the operational attributes of the model M. In this attempt, we rely upon our fundamental prior knowledge of the mechanisms involved, as well as upon our intuition. This is how a model is normally constructed. (2) *Prediction:* Knowing the expectable future pattern of inputs I, the present values of state variables S, and the operational attributes of the model M, we can now predict the future course of the response variables R. This is how a model is normally used for predictive purposes. (3) *Optimization:* Given I and some desired values of R, modify M so that I acting on M will result in R. This is how a model is sometimes used to guide optimization of controllable parameters within the system. If input factors I are controllable, we may also wish to seek their optimal values such as, acting on model system M, at a known initial state S, will produce the desired results R.

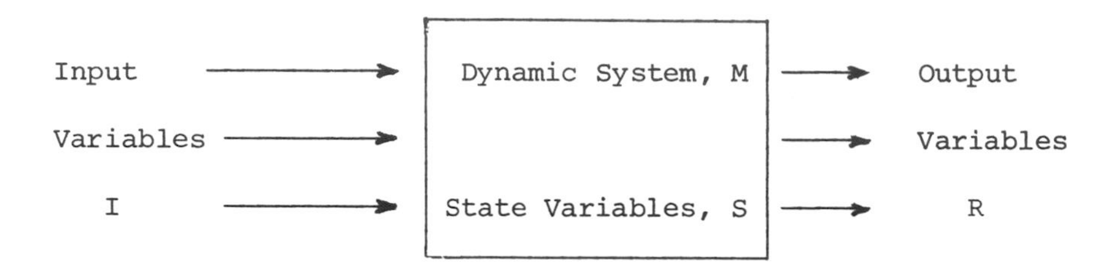

Figure 1: Conceptual representation of a dynamic state-dependent model.

F. Phases in the Construction and Operation of a Simulation Model

Several more-or-less distinct and sequential phases can be recognized in the course of the construction and operation of a simulation model, as follows: (1) Collection and analysis of real-world data representing phenomena pertinent to the subject of interest. (2) Definition of the problem which is to be answered by means of a simulation model. (3) Formation of a conceptual model. At this phase we focus upon the principal phenomena as they relate to each other quantitatively, and we conjecture as to the existence of functional relationships. (4) Formulation of a mathematical model based on the hypothesized functions. (5) Development of *algorithms*, which are logical, step-by-step procedures for solution of problems by manipulation of the mathematical model. (6) Check algorithms by making hypothetical calculations. (7) Programming of the mathematical model in an appropriate language for submission to a computer (using either a general-purposes language or a specialized simulation language). (8) Estimation of the necessary parameters of the system from the best available real-system data. If at all possible, special, independent experiments should be carried out to determine specific parameters. (9) Running-in the program to ensure that it is free of errors and that it is dimensionally and logically consistent within itself. (10) Testing the sensitivity of the model to various parameters and factors, particularly those which are uncertain, highly variable, or subject to our control in the real world. (11) Design and execution of simulation experiments pertinent to, and directed toward the solution of, the problem for which the simulation was undertaken in the first place. (12) Analysis of the simulation data and prediction of the real system's behaviour under various defined conditions. (13) Verification (or refutation) of the model by comparison of these predictions with independently obtained real

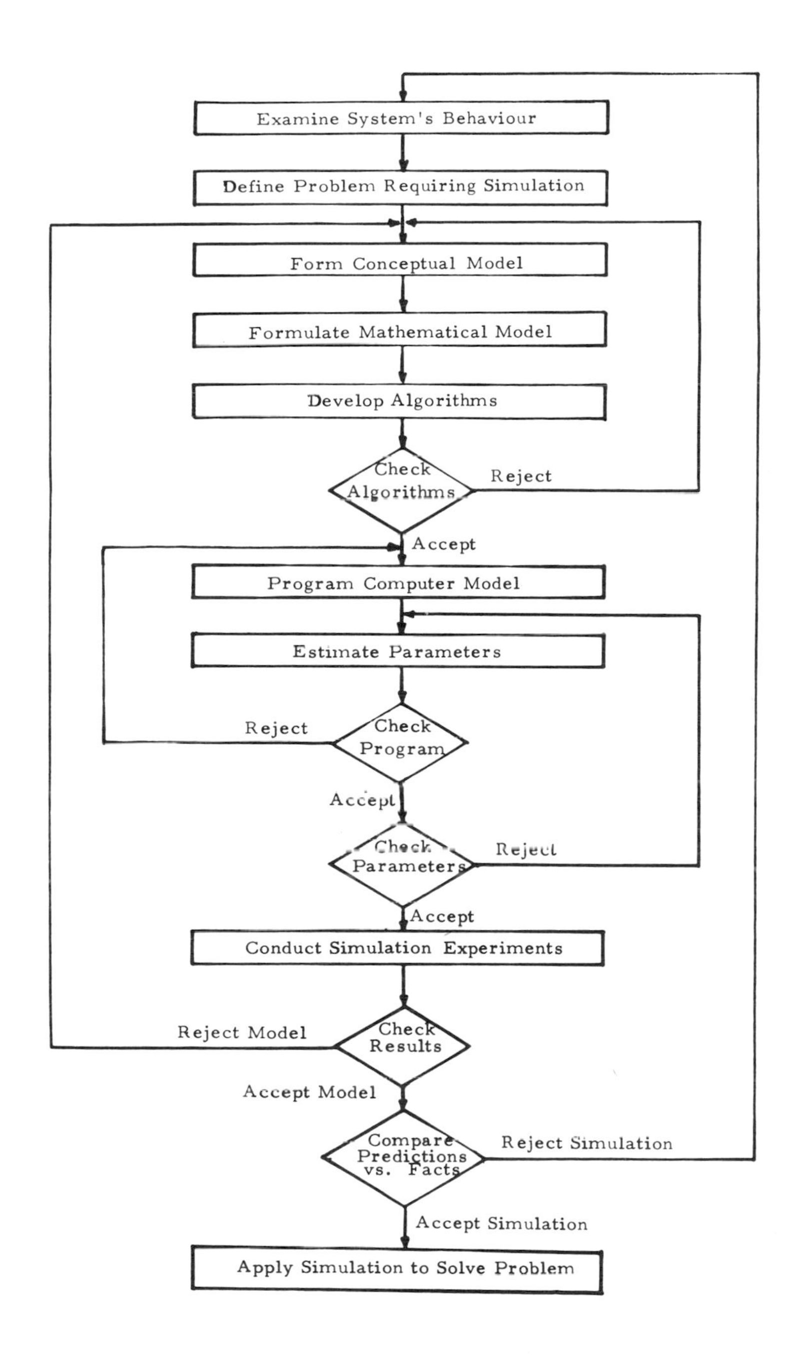

Figure 2: Phases in the construction of a simulation model.

world data. (14) Modification or further development of the model as deemed necessary to correct its flaws and improve its performance and predictive potential.

These steps make for a logical progression but obviously do not constitute a hard-and-fast rigid procedure. Rather, they are a tentative outline, or a "conceptual model," of model building. In practice, there are no sharp lines between the phases of a simulation, as the modeler interacts constantly with his model and takes such steps as he sees fit, sallying back and forth along the aforementioned sequence of steps, until a combination of trial-and-error, perseverance, ingenuity, proficiency, self-criticism, and good luck finally bring him to the point where he believes he has a satisfactory model.

If the model does not fit reality, we ask ourselves the following questions, among others: (1) Have we failed to account for an essential factor? (2) Have we confused the issue by including a needless, disturbing factor? (3) Have we erred in formulating the functional relations? (4) Have we erred in estimating the necessary parameters?

The construction of a simulation model is still as much an art as a science. The decision which relations to include in a model and which to exclude is very much a matter of judgment and of the degree to which one has a "feel" for the system. An important question is to establish the optimal size and degree of complexity of a model appropriate to the problem at hand. An overly comprehensive and complex model is too difficult to construct, analyze, and communicate. Its voluminous results may cloud the important issues in irrelevant detail. On the other hand, an overly simplified model may not be capable of exhibiting some effects or relationships which are essential in the operation of the system. A modeler must therefore seek a compromise between realism and ease of manipulation. The criteria for determining the appropriate size and level of detail of a model are partly logical, partly empirical, and largely pragmatic. There are no universal criteria, since so much depends on the purpose of the model. We must ask, first of all, what is the model for, what do we want it to do, and what are the proper scales of time and space, in each case.

There are in principle two general approaches to the construction and development of a simulation model for any complex system. The first is to start by a detailed modeling of subsystems, which can later be joined together, with appropriate attention to their interactions, to form a gradually more comprehensive model. The alternative is to start with a very general overall model of the entire system and later introduce greater detail in the treatment of the subsystems. In practice, one may find it necessary to apply both approaches to different portions of the model and thus go in both directions alternately or simultaneously until the desired levels of detail and comprehensiveness are achieved. These desired levels are likely to constitute a compromise between the requirements of realism on the one hand and of simplicity and ease of operation on the other.

G. Simulation Languages

An important problem is how to convert a mathematical model into a computer program that will, in turn, be used to carry out simulated experiments. In this process, we encounter two separate, and not necessarily compatible, criteria of efficiency. One pertains to the amount of time a research worker must initially invest in the preparation of the model. The second relates to the computer time eventually needed for running the model. Both criteria are involved in the selection of the appropriate programming language.

Clearly one way to approach programming simulation models is to write a special program for each system to be simulated in one of the well known general-purpose languages such as FORTRAN, ALGOL, COBOL, etc. This approach allows maximum flexibility in the design and formulation of the computational procedures, in the type and format of output records to be generated, and in the sort of simulation experiments to be performed with the model. FORTRAN is still the most widely used language for scientific applications. With FORTRAN, the programmer has the flexibility of being able to write almost any subroutine that he may need for a particular simulation program. Another advantage which FORTRAN has over the special purpose simulation languages is its almost universal applicability as FORTRAN compilers exist for almost every type of computer marketed today.

The shortcoming of this approach is the difficulty we often encounter in writing simulation programs in a general-purpose language if we are not expert programmers and do not have an expert available to do our bidding. A novice attempting to write a simulation program using a general-purpose language is always in danger of getting entangled in the complexities of sequencing the inter-related procedures forming the model. These technical problems are generally of no great interest in themselves, but afford fertile ground for minor errors which can be excruciatingly difficult to eradicate. Moreover, the more complex and detailed a program, the more difficult it becomes to communicate it to others, who, trying to decipher it may become so enmeshed in the thicket of branches they cannot see the trees, much less the forest as a whole.

However, recent years have witnessed the advent of several specialized simulation languages that are aimed at simplifying the task of writing simulation programs for a variety of different types of models. Among these languages have been the ones designated by the following acronyms: SIMSCRIPT, GASP, MIDAS, SIMPAC, MIMIC, DYNAMO, SIMULATE, and CSMP. These and other programs have been developed with the following objectives in mind: (1) To provide a generalized structure for designing simulation models; (2) to provide a rapid way to convert a simulation model into a computer program; (3) to facilitate making such changes or variations in a simulation model as may become necessary from time to time; (4) to provide a standardized yet flexible format for tabulating and graphing output for analysis. The simulation languages thus represent software systems designed especially for users who are not professional programmers. By allowing the research worker to concentrate on the problem and fundamental logic of the model rather than on the detailed procedure of computation, a simulation language can liberate the user from the tedium and distraction of programming techniques per se.

While the user's time is better spent, computer time may eventually be wasted through the use of a language containing subroutines not specifically adapted to any particular model. The question is, of course, which is the scarcer resource — the researcher's time or the computer's? It is a question for which there can be no universal answer.

The simulation languages available today differ considerably in the extent to which they are adapted to particular types of systems and in the extent to which they can render the programming of some simulation procedures more or less automatic. The decision on which language to use rests in each case upon such considerations as compatibility of the language with the available computer's software systems, cost of programming, availability of expert assistance, cost of computer time, and, not the least, upon the preference or experience of the modeler.

Ultimately, some of us would like to have a simulation language with just one instruction: "solve!" Until that blissful day arrives, however, there is still no substitute for a conscious effort to study the structure and techniques of simulation. The technical difficulties associated with simulation are only reduced, not eliminated, by the availability of simulation languages. Perhaps it is all for the best, for if simulation were made too easy, we might be tempted to apply it promiscuously without proper appreciation of its mode of operation, and hence of both its potentialities and its limitations.

H. Continuous Systems Modeling Program (CSMP)

Currently one of the most widely used and most versatile simulation languages is the Continuous Systems Modeling Program (CSMP). The first version of this language, designated "1130 CSMP" (for the computer series with which it was associated), was introduced by IBM in 1966. It was followed a year or two later by the more powerful and versatile "System 360 CSMP." The latter, in turn, was supplanted some years later by a still more sophisticated version known as "CSMP III." Basically, these programs were oriented toward a computer system with a core storage (memory) of at least 102 K bytes. According to the developers of CSMP (Brennan and Silberberg 1968), their main aim was to provide a "problem oriented" language for a user who is neither particularly proficient in computer programming nor especially interested in making the effort to acquire such proficiency. Simplicity of usage was therefore a prime design consideration. At the same time, however, the designers attempted to build a language which, basic simplicity notwithstanding, could yet be open-ended.

Some of the most significant features of CSMP are the algorithms it includes for automatically sorting the sequence of operations so as to achieve parallelism in calculations involving simultaneously occurring processes, the algorithms for the solution of implicit loops, the advanced integration formulas, the functional blocks, and the graphic output devices to help display and interpret the results of simulation trials. CSMP is entirely compatible with and uses the basic rules and conventions of FORTRAN. In fact, CSMP augments FORTRAN in a number of ways which are particularly useful in the simulation of dynamic systems. The system actually consists of a two-stage compiler, which first translates all input statements into FORTRAN and then into machine language for execution. This allows the program to incorporate all the features of FORTRAN, if the user so desires. Thanks to its built-in subroutines, CSMP makes it possible to write programs for complex systems in compact form, which is easy to read and communicate. It is particularly adapted to systems which can be described in terms of sets of algebraic and differential equations.

A complete and concise description of CSMP is given in the User's Manual distributed by IBM. Detailed explanations of the principles and procedures of CSMP have been provided by Brennan and Silberberg (1968), and by Brennan (1968). CSMP models for various ecological systems have been published by PUDOC in Holland (see the References section of this book). Specific programs for the simulation of soil physical processes will be elucidated in the following chapters. For the time being, therefore, we will merely point out in general terms some of the principal features of the CSMP language which we have found to be most useful in our work.

A CSMP model is normally segmented into three parts: INITIAL, DYNAMIC, and TERMINAL. All operations specified in the INITIAL part of the program are carried out prior to the actual simulation and are not repeated every time step. All operations in the DYNAMIC part are performed repeatedly and updated for each elapsed time-interval during the period of simulation. Finally, the TERMINAL part consists of calculations or decisions which are to be made only after the dynamic procedure has been completed. Segmenting of the program is not mandatory, however. When no segmenting is indicated (*i.e.*, in the absence of the INITIAL, DYNAMIC, and TERMINAL cards which normally divide the program into segments), the computer will assume that the entire program is DYNAMIC. A particularly useful feature of the language is the ease with which the program can be made, at the outset, to execute repeated simulations with alternative values of parameters, the results of which are graphed on the same scale for convenience of comparison.

Perhaps the single most powerful and important feature of CSMP is the statement which causes integration to be performed with respect to time. This statement has the general form:

Y = INTGRL (IC,X)

which calculates the output variable Y by integrating the differential function X, with the initial condition that Y at the start of the simulation is equal to IC (a specifiable constant). By means of the METHOD statement, a choice may be made among at least seven numerical integration techniques, two of which allow the integration interval (the time-step) to be varied and thus maximized to achieve greater computer-time efficiency. Once the appropriate integration method has been evoked, the process is carried out automatically by the program according to a specified error criterion. If a method is not specified, the program automatically chooses the fourth-order variable-step Runge-Kutta method.

Another very useful capability of CSMP is the computation of continuous functions from discrete tables. Thus, if paired values of two associated input variables are given in tabular form, interpolation between the given discrete values can be carried out either linearly, by calling upon AFGEN (the "arbitrary function generator"), or quadratically by using NLFGEN (the "non-linear function generator").

In a CSMP program, the statements may be arranged in any order. Unless specifically prevented from doing so, the compiler will automatically rearrange the statements in the appropriate sequence for efficient computation. This feature can be overruled by the simple insertion of a NOSORT card, which tells the computer that the following statements are not to be sorted but must be handled in the order in which they appear in the input program. The sorting routine can again be evoked by the single insertion of the statement SORT.

A group of statements forming a single functional element, when placed between the two cards PROCEDURE and ENDPRO, is treated as an entity in the sorting process. A "super block" of statements which constitute a subroutine to be invoked repeatedly during the simulation can be programmed most efficiently by the use of the MACRO designation, which is terminated by an ENDMAC card.

If a set of equations occurs which cannot be solved explicitly, the IMPLICIT loop procedure provides for a solution by iteration (a self-correcting, repeated trial and error method of numerical solution). The form of the IMPLICIT statement is:

Z = IMPL (IZ, ERROR, DUMZ)

wherein IZ is our initial guess for Z, ERROR is the allowable convergence error for the final value of Z calculated by the iteration process, and DUMZ is the dummy name used during the iteration for Z. DUMZ is checked against the guessed Z each time. A set of as many statements as necessary may be used to define Z, but the last statement of the loop must have DUMZ to the left of the equal sign. If, in the process of writing a program, we are unaware of the existence of an implicit loop, an error message will be provided by the CSMP system with a list of all the variables within the loop. An example of the use of the implicit loop solution is given in our section on the simultaneous calculation of the soil water and energy balances.

CSMP provides a number of convenient output formats. Perhaps the most useful is the PRTPLT (followed by a list of the desired variables), which causes the printer to plot the variables at specified time intervals (OUTDEL), with the value of each variable and the time of each plotted point being printed with the plot. The scale of the plot is set automatically by the system, but may also be specified at will by the programmer. The same goes for the label of each print-plot.

These are only a few of the convenient features of CSMP. Other features include methods of ending or aborting a simulation run when critical values are reached, and numerous built-in mathematical function blocks such as differentiation, delay functions, step functions, ramp functions, impulse generator, pulse generator, trigonometric functions, noise (random) generator, quantizer, dead-space, etc. In mentioning all this, we do not mean to imply that CSMP is necessarily the best simulation language available, only that we have found it convenient for our purposes at the present time.

I. Validation of Simulation Models

Without proof of validity, a model, however elegant, may be nothing but a tentative exercise in abstract logic. The problem of how to validate a simulation model remains, however, the most critical, difficult, and elusive of all problems associated with computer simulation.

Many people speak and write about "verification" of models. We believe this term may be misleading in the present context. Verification, meaning "proving the truth" of a model, involves a basic philosophical problem. To prove a model "true," we must be able to distinguish truth from falsehood. Yet the concept of truth has eluded philosophers throughout the history of thought. Since all models are approximations, none can be absolutely true, even if absolute truth does exist and is recognizable in our world (which is doubtful).

In the words of Reichenbach (1951): "Scientific philosophy refuses to accept any knowledge of the physical world as absolutely certain. Neither the individual occurrences, nor the laws controlling them, can be stated with certainty. The principles of logic and mathematics represent the only domain in which certainty is attainable; but these principles are analytic and empty. Certainty is inseparable from emptiness."

The truth of any model is at best partial, and hence relative; a model can only be "more-or-less" true. Moreover, a model can seem to be true in some ways, and far from it in other ways. In fact, what we are looking for is not "truth," or "verity," but *validity,* which is defined in Webster's Dictionary as "the quality of having force or being based on evidence or sound reasoning."

A model is scientifically valid if its assumptions conform to basic scientific principles. This is important, for if we build a model without a scientific basis then we are merely modeling our own ignorance. But internal scientific validity is not enough. A model must also be realistic. If a model is to portray a real system, then it must incorporate the major processes and phenomena which govern the system's behaviour. A model can be logically and scientifically valid within itself and yet fail to be realistic, simply because of the continual impact of factors disregarded in the analysis which may obscure the phenomena of direct interest to us.

We come to the crucial question of how well the predictions of a model or the data generated by it conform to independently obtained observations of the real system. We can better appreciate how difficult it is to answer that question if we bear in mind that these observations in themselves are only a partial and chance-affected sample of the real system and do not encapsulate its entire spectrum of variable phenomena.

In comparing model results to real-system data, one may take two approaches: comparison with historical records or forecasting of future events. If the historical approach is taken, one must be careful to avoid using the same set of data from which the model was estimated in the first place. Too many modelers, after building a model to fit a given set of data, turn around and announce triumphantly that their model exhibits "excellent fit" with the real system. This is circular thinking at its worst.

The ultimate test of a simulation model, in any case, is the accuracy with which it predicts the actual system (being simulated) in the future. Even here, one must be wary of the all too human tendency to select the data which fit the model best. (Modelers often develop a vested interest in the success of their creations and hence are in constant danger of losing their objectivity).

A valid model is one which does what it is supposed to do, which is to measure and predict the variables which are of interest to us with sufficient accuracy. The key words here are "sufficient accuracy." To establish that, one must acquire sufficient evidence, and judge the evidence objectively. The acquisition of sufficient evidence often requires a great expenditure of time and effort, and then may yield only data of a probabilistic nature. In fact, this task may be more difficult than that of developing a model, which is one reason why so many modelers, the author included, are so often tempted to publish their models without validation. On the other side of the argument, however, there is something to be said for the publication of models separately from their testing. Let a seemingly-plausible model be tested by someone other than, and totally independent of, the original modeler. The trouble is that we can seldom find scientists willing to test other scientists' models; nowadays most seem to be too busy constructing their own!

Since no model is expected to fit the data exactly, the question is whether the discrepancies are sufficiently small to be tolerable. Also, the question is if the errors are random or systematic. The fit and forecasting ability of the model may be better for some variables than for others. Furthermore, in a dynamic model, it may happen that the calculated values gradually suffer from accumulating errors and the forecast worsens progressively. Hence a dynamic model may be valid enough for a limited time period but not indefinitely.

There appear to be at least three ways in which the validation problem in process models can be approached (Cohen and Cynert, 1961). First, statistical methods can be used to test whether the actual and model-generated time series display similar timing and amplitude characteristics. Second, simple regressions of the generated series as functions of the actual series can be compared, and then we can test whether the resulting regressions have intercepts which are not significantly different from zero and slopes which are not significantly different from unity. Third, we could perform a factor analysis on the set of generated time paths and a second factor analysis on the set of observed time paths, and then we can test whether the two groups of factor loadings are significantly different from each other.

In practice, such rigorous testing is seldom performed. Yet the confidence with which we can use simulation results to guide us in understanding and managing the real system is proportional to how well we have been able to validate our model.

By way of summary, we can refer to a statement by Zeigler (1976) concerning the degrees of strength for model validity. A model is *replicatively valid* if it matches data already acquired from the real system. Stronger yet is a model which is *predictively valid* in the sense that it can match data *before* they are acquired. Finally, the highest level of validity is that which characterizes a *structurally valid* model, *i.e.*, which not only matches the results obtained from experiments or observations but also reproduces the structure and mechanisms of the real system that produces those results. In science, we aim toward structurally and mechanistically valid models which reveal the workings of nature.

J. Conclusions from Simulation Experiments.

When the results of a more-or-less validated model are analyzed, evaluated, and summarized, we can draw certain limited conclusions from our simulation experiments. These conclusions must be based upon the systematic testing of explicit hypotheses over a realistic range of values of input variables and parameters. The conclusions should be stated explicitly and substantiated by the model results.

An example of the sort of conclusions which we can perhaps draw from a simulation experiment is the following statement: "We conclude from a series of tests with a mechanistic model of soil water and energy dynamics that a treatment causing the whitening of the soil surface (thereby increasing the albedo to 50%) can reduce cumulative evaporation from a medium-textured soil by about 20% during the first week of a post-irrigation dry period in a semi-arid climate. Thereafter, the effect can be expected to decrease gradually and in effect to disappear after about one month. We surmise that this practice may have a beneficial effect on germination and seedling establishment where these processes may be limited by soil moisture but not by soil temperature. The practice is not likely to be worthwhile as a method of soil water conservation in the long run." (The reader may wish to compare this statement to the results of our actual simulation experiment reported in Chapter 2 of this monograph.)

We note that to the very end the conclusions of a simulation model remain somewhat tentative and quantitatively uncertain. Yet, such conclusions give us a yardstick by which to evaluate existing practices and to recommend improvements. Obviously, our conclusions themselves bear further testing, research, and re-research, as do our models. And so *ad infinitum*.

K. Communication of Models and Simulations

Even after construction, validating, and experimenting with a simulation model, our effort is likely to come to naught if its accomplishment is not properly communicated. In the words of Zeigler (1976), "what posterity will remember is not the intellectual and emotional gratification accrued to the modeler (though this is what he may remember). The long term contribution of any modeling effort lies in the benefits it affords, either by direct use or by guidance for further development, to science and industry."

Various modes or stages exist for communication of models: (1) Informal description of the conceptual model and the assumptions upon which it was based. (2) Formal presentation of the mathematical model. (3) Detailed description of the computer program designed to implement the model. (4) Elucidation of the experiments performed and analysis of their results. (5) Conclusions about the significance, applicability, limitations and costs of using the model for various purposes. (6) Comparison of the model to alternative models and projection of further work.

The mode or modes of communication to adopt in any set of circumstances depends to a considerable degree on the type of audience, *i.e.*, on whether one aims to transmit the model and its results to potential users of the model as is, or to fellow scientists who may wish to make indirect use of it in their own investigations. Too many models devised in recent years have gone to waste simply because they were never communicated, or communicated in an ineffective way, to its potential users or other parties who may be interested in it for one reason or another. Failure to describe the model effectively can even boomerang against the modeler himself, who may wish to return to his own model after some time away from a project, only to discover that he has forgotten essential features which he had once taken for granted as self-evident and hence failed to record. For these reasons, the task of modeling is never complete if it does not culminate in a detailed, well-annotated, and well-documented and readable program description.

In this monograph, we attempt to apply these criteria in communicating several of our recent models to colleagues engaged in studies of the soil-water-plant system.

I. ISOTHERMAL EVAPORATION OF SOIL WATER UNDER FLUCTUATING EVAPORATIVITY, INCLUDING THE ROLE OF HYSTERESIS

A. Background and Description of the Problem

Evaporation of soil moisture can occur through extraction and transpiration by plants, or directly from the soil surface. When the two processes are concurrent and inseparable, they are generally treated as a single process called *evapotranspiration*. When the surface is covered entirely by active vegetation, extraction by roots and transpiration by foliage predominate over direct evaporation. This case will be considered in Chapter IV. In the total absence of vegetation, however, evaporation takes place entirely from the soil. This process is the subject of our present chapter and of the one immediately following it.

The process of evaporation, if uncontrolled, can involve very considerable losses of water in both irrigated and rainfed agriculture. Under annual field crops, the soil surface normally remains bare through the periods of tillage, planting, germination, and early seedling growth, periods in which evaporation can deplete the moisture of the surface soil and thus affect the growth of young plants during their most vulnerable stage. Rapid drying of a seedbed can doom an entire crop from the outset. The problem is also encountered in young orchards, where the soil surface remains largely bare for several years. Finally, the problem is most acute in dryland farming, where the land is regularly fallowed for a number of months to collect and conserve rainwater from one season to the next. In semi-arid regions, evaporation has often been found to cause the loss of over half the seasonal amount of rainfall, which might otherwise be stored in the soil and be available for subsequent crop use. It is therefore important to be able to predict the rate and cumulative amount of water loss, as well as the distribution of moisture within the soil and in the seed zone during the evaporation process.

Three conditions are necessary to sustain evaporation from a porous body. Firstly, heat must be supplied to meet the latent heat requirement of evaporation. Secondly, the vapour must be transported away from the zone of evaporation, by diffusions or convection or both. Thirdly, there must be a continual supply of water from the interior of the body to the evaporation site. The first two conditions can be considered to be external to the porous body, as they are influenced by such meteorological factors as radiation, air temperature and humidity, and wind velocity, which together determine the *atmospheric evaporativity*. The third condition depends upon the content and potential of water in the porous body and upon its conductive properties, which determine the maximal rate at which the body can transmit water to the evaporation site. Accordingly, the evaporation rate of soil moisture is limited either by external evaporativity or by the soil's own ability to deliver water, whichever is the lesser at any time.

Among the various sets of conditions under which evaporation may occur are the following: (1) A water table may be present at a constant or variable depth near the soil surface, or it may be absent or too deep to affect evaporation, in which case continued evaporation causes the soil to dry out. (2) The soil profile may be uniform with respect to its basic properties, or it may exhibit gradual or abrupt variations with depth, such as distinct layers differing in texture or structure. (3) The structure of the soil matrix may be stable, or it may swell and shrink as the soil alternately wets and dries. As the surface zone tends to dry more rapidly than the interior, differences in shrinkage can cause the soil to break up into aggregates, or to form cracks, which, in turn, constitute secondary drying planes. (4) The flow pattern may be one-dimensional (vertical) or two- or three-dimensional, as in the presence of vertical cracks or slanted layers. (5) Conditions may be isothermal, or nearly so, or strongly heterothermal. In the latter case, thermal gradients may induce coupled heat and vapour flow within the soil. (6) External environmental conditions may remain constant or fluctuate either randomly or cyclically. (7) Soil moisture flow may be governed by evaporation alone, or by both evaporation at the surface and drainage at the bottom of the profile.

The case of non-isothermal evaporation is treated in our next chapter, whilst the present one considers isothermal evaporation from a stable, uniform, one-dimensional soil profile, in the absence of a water table, under constant and diurnally-fluctuating evaporativity.

B. Previous Studies

The process by which soil moisture evaporates and the soil surface dries has been studied by numerous investigators during the last two decades (*e.g.*, Hide 1954; Lemon 1956; Gardner 1959; Wiegand and Taylor 1961; Gardner and Hillel 1962). Many treatments of this process (*e.g.*, Black *et al.* 1969; Gardner 1973) were based on the concept that the soil surface is subjected to a constant, meteorologically induced, *evaporativity* (or *potential evaporation)*, which is generally conceived to be the maximal, and hence, limiting, rate of evaporation possible from the soil as long as its surface is kept sufficiently wet (Hillel 1971).

Under constant evaporativity, the evaporation process can be divided into two, and possibly three, stages (Philip 1957, 1967; Feodoroff and Rafi 1962): a *constant-rate stage*, controlled by external evaporativity; a *falling-rate stage*, controlled by the soil profile's transmission of water to the evaporation zone; and a *vapour diffusion stage*, during which evaporation continues at a very slow and relatively constant rate controlled by the vapour diffusivity of the dried surface zone.

The end of the first stage has generally been assumed to occur whenever the soil surface has been desiccated to a point of "air-dryness" (Hillel 1971). This is admittedly an approximation, but the exact value of the critical soil surface wetness at which the transition takes place is probably not critical, in any case.

In actual nature, evaporativity is obviously not constant but intermittent, as it fluctuates diurnally and varies from day to day, so it may become difficult or even impossible to discern or distinguish between the stages described above. The resulting course of evaporation may not be described accurately by a simplistic theory based on the assumption of constant evaporativity.

Recently, detailed experimental observations by Jackson (1973) and Jackson *et al.* (1973) showed that the surface-zone soil moisture content fluctuates in a manner corresponding to the diurnal fluctuation of evaporativity; that is to say, the soil surface dries during daytime and tends to rewet during nighttime, apparently by sorption from the moister layers beneath. This pattern was found to occur in a layer of soil several centimetres thick. It remains to be established to what extent this interesting and possibly significant phenomenon can be predicted *a priori* on the basis of soil moisture flow theory, and how the magnitude of this fluctuation can affect the cumulative amount of evaporation over a period of several days or weeks.

To account for the effect of varying meteorological conditions on evaporation dynamics, it seems useful to construct a simulation model capable of monitoring the process continuously through repeated cycles of increasing and decreasing evaporativity. Such a model might clarify the extent to which the diurnal pattern of evaporativity, heretofore generally ignored, could influence the overall quantity of evaporation and the moisture distribution in space and time.

Numerical models of soil moisture evaporation have been published by Hanks and Gardner (1965), Hanks *et al.* (1969), Ripple *et al.* (1972), and Van Keulen and Hillel (1974). The model presented herein is based on recently published simulation studies by the author (Hillel, 1975, 1976).

C. Governing Equations

The flow of an incompressible fluid in a rigid, homogeneous, isotropic, and isothermal porous medium can be described by a combination of two equations:

(1) *Darcy's law* which states that the flux of water (q) is proportional to, and in the direction of, the driving force which is the effective potential gradient.

$$q = -K\nabla\phi \qquad (1.1)$$

where ϕ, the hydraulic potential, is the algebraic sum of the matric potential (ψ) and the gravitational potential. Expressed in head units (free energy per unit weight), the hydraulic potential can be written as

$$\phi = \psi - z \qquad (1.2)$$

where z is the gravitational level expressed as depth below the soil surface. K is the hydraulic conductivity which in an unsaturated soil is a function of the water content, θ.

(2) *The continuity equation,* which states that the time (t) rate of change of water content in a volume element of soil must equal the divergence of the flux (q)

$$\frac{\partial\theta}{\partial t} = -\nabla . q \qquad (1.3)$$

These two relations are combined to give

$$\frac{\partial\theta}{\partial t} = \nabla . (K\nabla\phi) \qquad (1.4)$$

which, in one-dimensional form, becomes

$$\frac{\partial\theta}{\partial t} = \frac{\partial}{\partial x}(K\frac{\partial\phi}{\partial x}) \qquad (1.5)$$

If the system considered is vertical, and the z direction is taken as positive from the soil surface downward, we obtain

$$\frac{\partial\theta}{\partial t} = \frac{\partial}{\partial z}[K\frac{\partial(\psi - z)}{\partial z}] = \frac{\partial}{\partial z}(K\frac{\partial\psi}{\partial z}) - \frac{\partial K}{\partial z} \qquad (1.6)$$

For monodirectional upward movement of water from an initially uniformly-wet profile toward an evaporation zone at the soil surface, equation (1.6) is solved subject to the following initial and boundary conditions:

$$
\begin{array}{llll}
t = 0, & z \geq 0, & \theta = \theta_i & \\
t > 0, & z = z_b, & q = 0 & \\
t > 0, & z = 0, & \theta > \theta_d, & q = E_\circ \\
t > 0, & z = 0, & \theta \leq \theta_d, & q = -K(\theta)\dfrac{\partial \phi}{\partial z}
\end{array}
\qquad (1.7)
$$

wherein t is time, θ volumetric wetness, θ_i the soil's initial wetness, z depth, z_b the bottom of the profile, θ_d the air-dry value of the soil at the surface and $E_\circ$ the climatically imposed evaporational demand (evaporativity), which in our simulation was assumed to vary diurnally following a sine function during daytime, as follows:

$$E_\circ = E_{max}\sin(2\pi t/86400) \qquad (1.8)$$

where E_{max} is the maximal midday evaporativity and t is time in seconds from sunrise.

To convert this mathematical model into a form soluble by a digital computer, the differential equations of water transport in the soil are cast into explicit algebraic equations, involving the values of the variables as they exist at discrete points in space and time. This will be shown in the computer program which follows.

D. Description of the Computer Model

The geometric structure of the model is shown in Figure 1.1 which depicts a uniform soil profile of finite depth divided into NL compartments, not necessarily of equal thickness (TCOM). The rate of water movement (FLUX) between compartments obeys Darcy's law in finite-difference form. The wetness of a compartment at any moment of time determines the compartment's matric potential and hydraulic conductivity.

The actual program, written in System 360 CSMP (IBM 1972) is presented in Figure 1.2. Apart from the formal STORAGE, DIMENSION, EQUIVALENCE, and FIXED specifications given at the start, the program consists of three segments (as explained in section H of our introductory chapter): INITIAL, DYNAMIC, and TERMINAL. (Note: CSMP statements obey FORTRAN conventions).

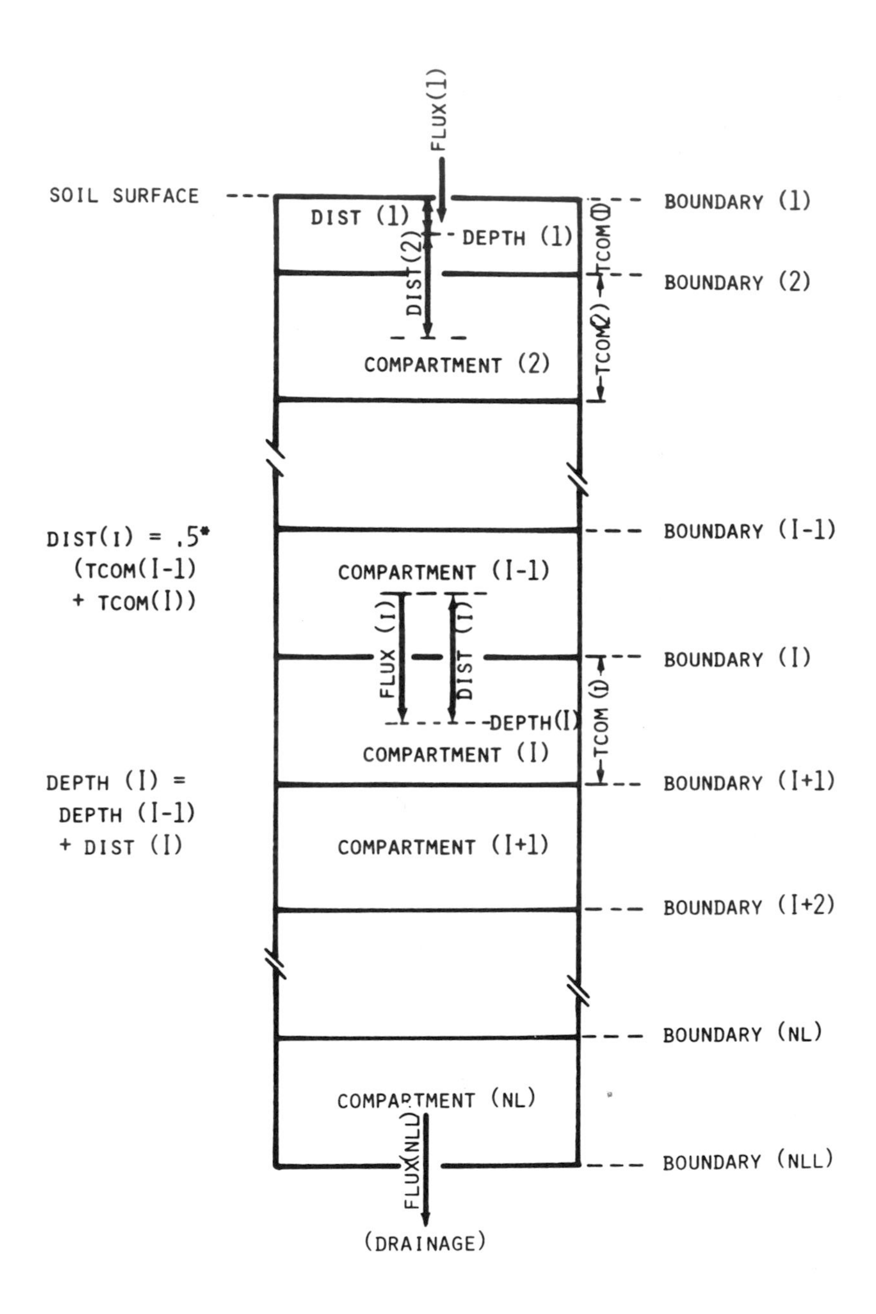

Figure 1.1. Geometric scheme used in simulating flow in a one-dimensional vertical soil profile.

The INITIAL Section

This section begins with the specification NOSORT, requiring the computer to perform operations in the sequence listed. The number of compartments, NL, is arbitrarily set at 14, with thicknesses ranging from 0.01 m at the top to 0.10 m at the bottom, as listed in TABLE TCOM. The total profile depth is 1 m.

The initial volumetric wetness of each compartment, as given in TABLE ITHETA, is 35 percent. The product of ITHETA and the compartment thickness then gives the volume of water per unit area of soil surface, IVOLW (in depth units), initially contained in each compartment. The DEPTH of each compartment is the vertical distance between its midpoint and the soil surface, while DIST is the flow path length from the midpoint of any compartment to the midpoint of the adjacent one.

The total daily potential evaporation was set at 0.01(10mm). Thus, the steady evaporativity, AVPET, was taken to be 0.01/86400 m per sec. This was compared to a sinusoidally fluctuating pattern of evaporativity having an amplitude (AMP) of π*AVPET. The "air-dry" state of the soil surface compartment, corresponding to its lowest possible wetness, is specified in terms of its minimal matric potential head, MINPOT (in meters).

FUNCTION SUCTB is a table of volumetric wetness versus suction, the latter being the matric potential head (*m*) taken as positive. At this stage, no account is taken as yet of hysteresis in the soil moisture characteristic. FUNCTION CONDTB is a table of volumetric wetness versus hydraulic conductivity. The values used for these functions pertain to Gilat fine sandy loam, a loessial soil occurring in the Northern Negev of Israel.

The DYNAMIC Section

The following calculations are made and updated at each time step during the simulation:

(1) The volume of water in each of the 14 compartments is the time-integral of the net flux NFLUX, which is the difference between the influx and outflux of each time-step:

```
VOLW1 = INTGRL(IVOLW1, NFLUX1,14)                    (1.9)
```

(2) Volumetric wetness (THETA) of each compartment is the ratio of the water volume to the compartment volume per unit area (TCOM):

```
THETA(I) = VOLW(I)/TCOM(I)                           (1.10)
```

(3) Matric potential is read from the suction table:

Figure 1.2. CSMP listing for calculating moisture distribution and evaporation from bare soil under steady and fluctuating evaporativity.

```
TITLE                    ISOTHERMAL EVAPORATION UNDER
*                    DIURNALLY-FLUCTUATING EVAPORATIVITY

*             UNITS
*         KG = KILOGRAMS
*         M  = METERS
*         S  = SECONDS

*                           GLOSSARY OF SYMBOLS

* AMP    = AMPLITUDE OF DAILY WAVE OF POTENTIAL EVAPORATION (M/S)
* AVCOND = AVERAGE HYDRAULIC CONDUCTIVITY FOR FLOW BETWEEN COM-
*          PARTMENTS (M/S)
* AVPET  = TIME-AVERAGE VALUE OF POTENTIAL EVAPORATION RATE (M/S)
* COND   = HYDRAULIC CONDUCTIVITY (M/S)
* CONDTB = CONDUCTIVITY TABLE (THETA VERSUS COND)
* CUMEVP = CUMULATIVE EVAPORATION (M)
* CUMPET = CUMULATIVE POTENTIAL EVAPORATION (M)
* CUMWTR = TOTAL WATER CONTENT OF PROFILE (M)
* DEPTH  = DEPTH OF MIDPOINT OF COMPARTMENT (M)
* DIST   = DISTANCE OF FLOW BETWEEN ADJACENT COMPARTMENTS (M)
* EVAP   = EVAPORATION RATE (M/S)
* FLUX   = FLOW RATE OF WATER (M/S)
* HPOT   = HYDRAULIC POTENTIAL HEAD (M)
* I      = INDEX OF COMPARTMENT (ORDINAL NUMBER)
* ITHETA = INITIAL VOLUMETRIC WETNESS (M3/M3)
* IVOLW  = INITIAL VOLUME OF WATER IN EACH COMPARTMENT (M)
* MINPOT = MINIMAL MATRIC POTENTIAL OF SURFACE COMPARTMENT (M)
* MPOT   = MATRIC POTENTIAL HEAD OF SOIL MOISTURE (M)
* NFLUX  = NET FLUX OF WATER INTO COMPARTMENT (M/S)
* NL     = NUMBER OF COMPARTMENTS COMPRISING THE PROFILE
* SUCTB  = SUCTION TABLE (THETA VERSUS -MPOT)
* TCOM   = THICKNESS OF COMPARTMENT (M)
* THETA  = WETNESS OF SOIL IN EACH COMPARTMENT (VOLUME FRACTION,
*          (M3/M3)
* VOLW   = VOLUME OF WATER IN EACH COMPARTMENT (M)


STORAGE        TCOM(25),ITHETA(25),DEPTH(25),DIST(25),COND(25)
STORAGE        AVCOND(25),FLUX(25),MPOT(25),HPOT(25)
/      DIMENSION  VOLW(25),IVOLW(25),THETA(25),NFLUX(25)
/      EQUIVALENCE  (VOLW1,VOLW(1)),(IVOLW1,IVOLW(1)),(NFLUX1,NFLUX(1)
FIXED          I,NL,NLL
*      UNITS   SI(MKS)

INITIAL

NOSORT
PARAMETER      NL=14, MINPOT=-1000.
               NLL=NL+1
               PI=3.14159
               AVPET=.01/86400.
               AMP=PI*AVPET
TABLE TCOM(1-14)=.01,.02,.03,.04,2*.05,8*.10
```

```
TABLE ITHETA(1-14)=14*.35
              IWATER=0.
        DO 100 I=1,NL
              NFLUX(I)=0.
              IVOLW(I)=ITHETA(I)*TCOM(I)
          100 IWATER=IWATER*IVOLW(I)
              DEPTH(1)=.5*TCOM(1)
              DIST(1)=DEPTH(1)
        DO 110 I=2,NL
              DEPTH(I)=DEPTH(I-1)+.5*(TCOM(I-1)+TCOM(I))
          110 DIST(I)=.5*(TCOM(I-1)+TCOM(I))
FUNCTION SUCTB=(.005,10000.),(.01,3500),(.025,1000.),(.05,200.),...
              (.1,40.),(.15,10.),(.2,6.),(.25,3.5),(.3,2.2),      ...
              (.35,1.4),(.4,.56),(.45,0.),(1.,-100.)
FUNCTION CONDTB=(.005,.4E-13),(.05,.5E-12),(.1,.15E-11),(.15,.8E-11),...
              (.2,.5E-10),(.25,.3E-9),(.3,.2E-8),(.35,.12E-7),          ...
              (.4,.8E-7),(.45,.5E-6),(1.,.5E-6)

DYNAMIC

NOSORT
              VOLW1=INTGRL(IVOLW1,NFLUX1,14)
        DO 200 I=1,NL
              THETA(I)=VOLW(I)/TCOM(I)
              COND(I)=AFGEN(CONDTB,THETA(I))
              MPOT(I)=-AFGEN(SUCTB,THETA(I))
          200 HPOT(I)=MPOT(I)-DEPTH(I)
        DO 210 I=2,NL
          210 AVCOND(I)=(COND(I-1)*TCOM(I-1)+COND(I)*TCOM(I))/2.*DIST(I)
              FLUX(NLL)=0.
        DO 220 I=2,NL
          220 FLUX(I)=(HPOT(I-1)-HPOT(I))*AVCOND(I)/DIST(I)
              PET=AMAX1(0.,AMP*SIN(2.*PI*TIME/86400.))
              IF (MPOT(1).GT.MINPOT) EVAP=PET
              IF (MPOT(1).LE.MINPOT) EVAP=AMIN1(PET,-FLUX(2))
              FLUX(1)=-EVAP
        DO 230 I=1,NL
          230 NFLUX(I)=FLUX(I)-FLUX(I-1)
              CUMEVP=INTGRL(0.,EVAP)
              CUMPET=INTGRL(0.,PET)
              CUMWTR=0.
        DO 300 I=1,NL
          300 CUMWTR=CUMWTR+VOLW(I)


TERMINAL

        TIMER  FINTIM=864000., OUTDEL=21600.
        PRINT  (optional)
        PRTPLT EVAP,PET,CUMEVP,CUMPET,CUMWTR
        METHOD RKS
        END
        STOP
```

```
MPOT(I) = -AFGEN(SUCTB, THETA(I))                          (1.11)
```

where AFGEN designates the arbitrary function generator of CSMP for tabular pairs of x, y coordinates.

(4) Hydraulic potential is obtained by summing the matric potential and the gravity potential (-DEPTH):

```
HPOT(I) = MPOT(I)-DEPTH(I)                                 (1.12)
```

(5) Average hydraulic conductivity for flow through boundary (*I*) between adjoining compartments (I) and (I-1) is weighted according to their thicknesses:

```
AVCOND(I)=(COND(I-1)*TCOM(I-1)+COND(I)*TCOM(I))/2.*DIST(I)
                                                           (1.13)
```

(6) The bottom boundary is taken to be an impervious plane:

```
FLUX(NLL) = 0.0                                            (1.14)
```

(7) Flux between compartments follows Darcy's law in discrete form:

```
FLUX(I) = (HPOT(I-1)-HPOT(I))*AVCOND(I)/DIST(I)            (1.15)
```

(8) Diurnally fluctuating potential evaporation is simulated as a sine function of time:

```
PET = AMAX1(0.0, AMP*SIN(2.*PI*TIME/86400.))               (1.16)
```

where AMP is π times the average evaporativity AVPET, as defined in the INITIAL section. Use of the AMAX1 specification prevents the potential evaporation rate from becoming negative and sets the nighttime evaporation rate at zero. In principle, the nighttime evaporation rate can be set at any other finite value or fraction of AVPET. Steady evaporativity can simply be simulated with the alternative statement:

```
PET = AVPET                                                (1.17)
```

(9) The actual evaporation rate is set equal to the potential rate as long as the matric potential of the topmost compartment remains greater than the initially specified air-dryness value, MINPOT:

```
IF (MPOT(1).GT.MINPOT) EVAP = PET                          (1.18)
```

After the topmost compartment drops to its minimum (air-dry) matric potential, or below, it can dry out no more and the evaporation rate becomes equal to the rate of upward transmission of moisture by the profile (or to the potential rate, whichever is the lesser):

```
IF (MPOT(1).LE.MINPOT)EVAP = AMIN1(PET,-FLUX(2))
```
(1.19)

In either case,

```
FLUX(1) = -EVAP
```
(1.20)

where FLUX(1) is the flux of water through the soil surface, and where the upward direction is considered as negative.

(10) The change in water content of each compartment obeys the continuity equation:

```
NFLUX(I) = FLUX(I)-FLUX(I-1)
```
(1.21)

where NFLUX stands for "net flux" and enters into the determinations of water volume and wetness values according to Eqs. (1.9) and (1.10).

The remainder of the program deals with the output and specifies the format by which the results are to be presented. The TERMINAL section includes statements of the total simulation time to be run (FINTIM), the time intervals for output (OUTDEL), and a list of variables to be printed or print-plotted. Finally, METHOD specifies the method of integration, and RKS stands for the Runge-Kutta fourth-order, variable time step, procedure.

E. Results of a 10-Day Simulation of Evaporation Without Hysteresis.

To illustrate the workings of the model, we carried out a 10-day simulation of evaporation under constant and under diurnally fluctuating evaporativity. The results are shown in Figures 1.3 through 1.6.

Figure 1.3 shows the cumulative values of potential evaporation (evaporativity) and of actual evaporation under both steady and diurnally fluctuating conditions. It is seen that the two evaporativity regimes being compared were equivalent in that they resulted in the same cumulative values of potential evaporation at the end of each day. In both cases, actual evaporation at first followed the course of the potential evaporation, but began to deviate from the latter during the second day and fell progressively below it during subsequent days. At the end of the 10-day period simulated, total actual evaporation under both regimes was only about 40 percent of the total potential evaporation. Between the end of second and the end of the tenth day, in fact, the actual evaporation was only about 25 percent of the potential. This was apparently a consequence of the desiccation

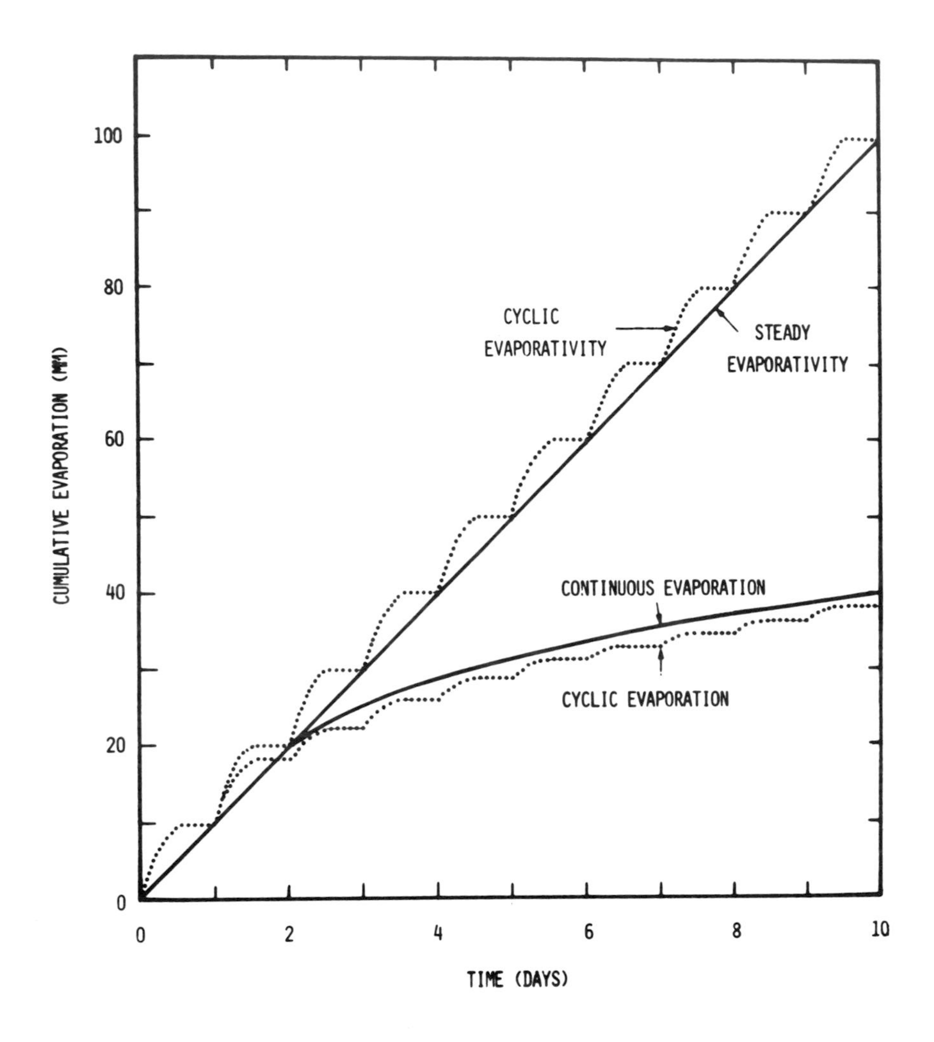

Figure 1.3. Cumulative values of potential evaporation (evaporativity) and of actual evaporation under steady and diurnally fluctuating conditions.

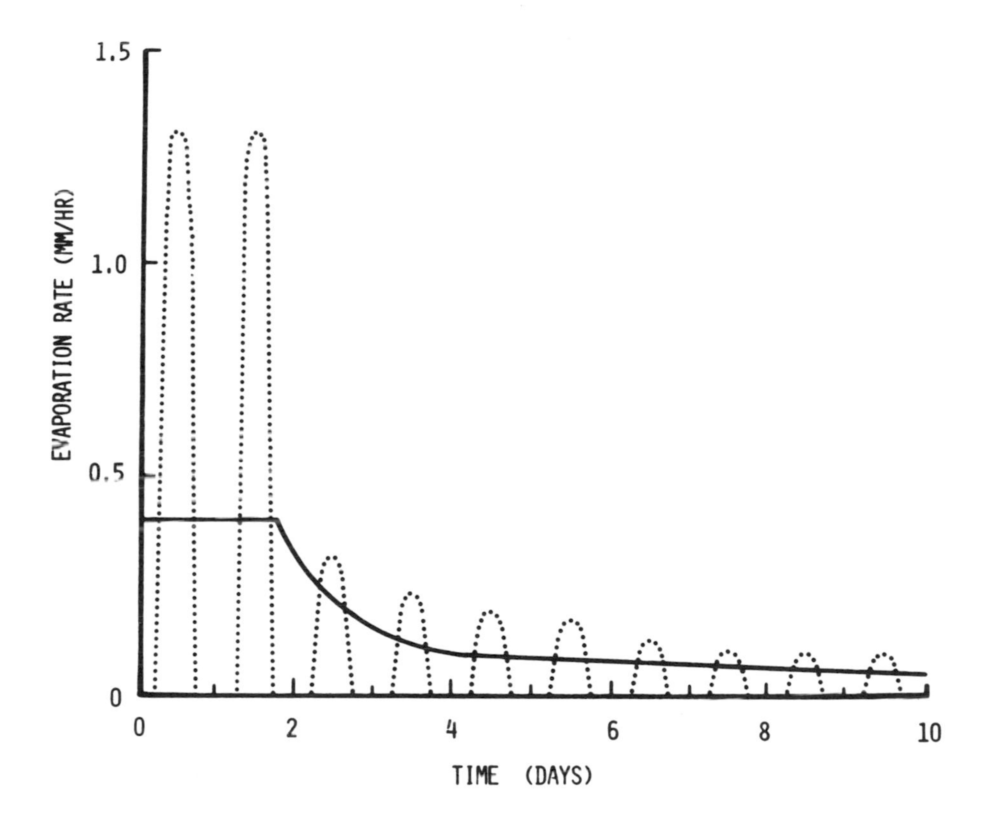

Figure 1.4. Evaporation rates under steady evaporativity (smooth curve) and under cyclic evaporativity (dotted curves).

of the surface zone which had occurred by the end of the second day, as seen in Figure 1.5.

A comparison between the 10-day cumulative evaporation value for the continuous versus cyclic evaporation regimes indicates that the latter resulted in a somewhat reduced total water loss.

Figure 1.4 presents the actual evaporation rates resulting from the same two evaporation regimes. Once again it can be seen that the actual evaporation rates equalled the potential rates during the first two days but fell off rapidly afterwards. The evaporation rate under the steady evaporativity, represented by the solid line curve, exhibited the classical three-stage pattern: an initial stage of constant, high-rate evaporation, followed by an intermediate stage of falling-rate evaporation, which finally culminated in a third stage characterized by a low and nearly constant rate of evaporation. On the other hand, the cyclic evaporation regime merely resulted in a series of daily wavelike evaporation phases which tended to damp down in time as the surface zone desiccated to progressively greater depth. Even after 10 days, however, these evaporation-time waves were still prominent, with the mid-day evaporation rate exceeding 1 mm/hr.

Figures 1.5A and 1.5B indicate the change in volumetric wetness of the soil at depths of 5 and 20 mm, respectively. Under steady evaporation, the surface zone is seen to fall very steeply to a value of air-dryness (about 2.4 percent moisture), which prevails after the second day. Under the cyclic evaporation, on the other hand, the surface-zone soil moisture content exhibits a wave-pattern, as the soil repeatedly dries down in daytime only to resorb moisture during the nighttime pauses in evaporativity. The net result of this effect is to increase the average moisture of the surface zone from about 2.4 percent to about 4 percent. While the magnitude (or amplitude) of these diurnal fluctuations in surface-zone wetness tends to decay in time, it is seen to be still very considerable at the end of the simulated 10-day period, with the daytime value dropping below 2.4 percent and the nighttime value rising above 5.5 percent. The corresponding moisture content fluctuations were progressively smaller in the deeper layers, and exhibited an increasing time lag, as might be expected.

The distribution of moisture in the entire profile at the end of the 10-day simulation is shown in Figure 1.6. Comparison of the curve for the soil subjected to cyclic evaporativity to the one for the soil under steady evaporativity reveals no differences beyond a depth of 0.1 m, but a widening divergence in the upper zone toward the soil surface. The area between the curves represents the increment of water evaporated under steady evaporativity as compared to the cyclic evaporativity regime. The accumulation of water at the bottom of the profile is a consequence of the bottom boundary condition which precluded through-drainage.

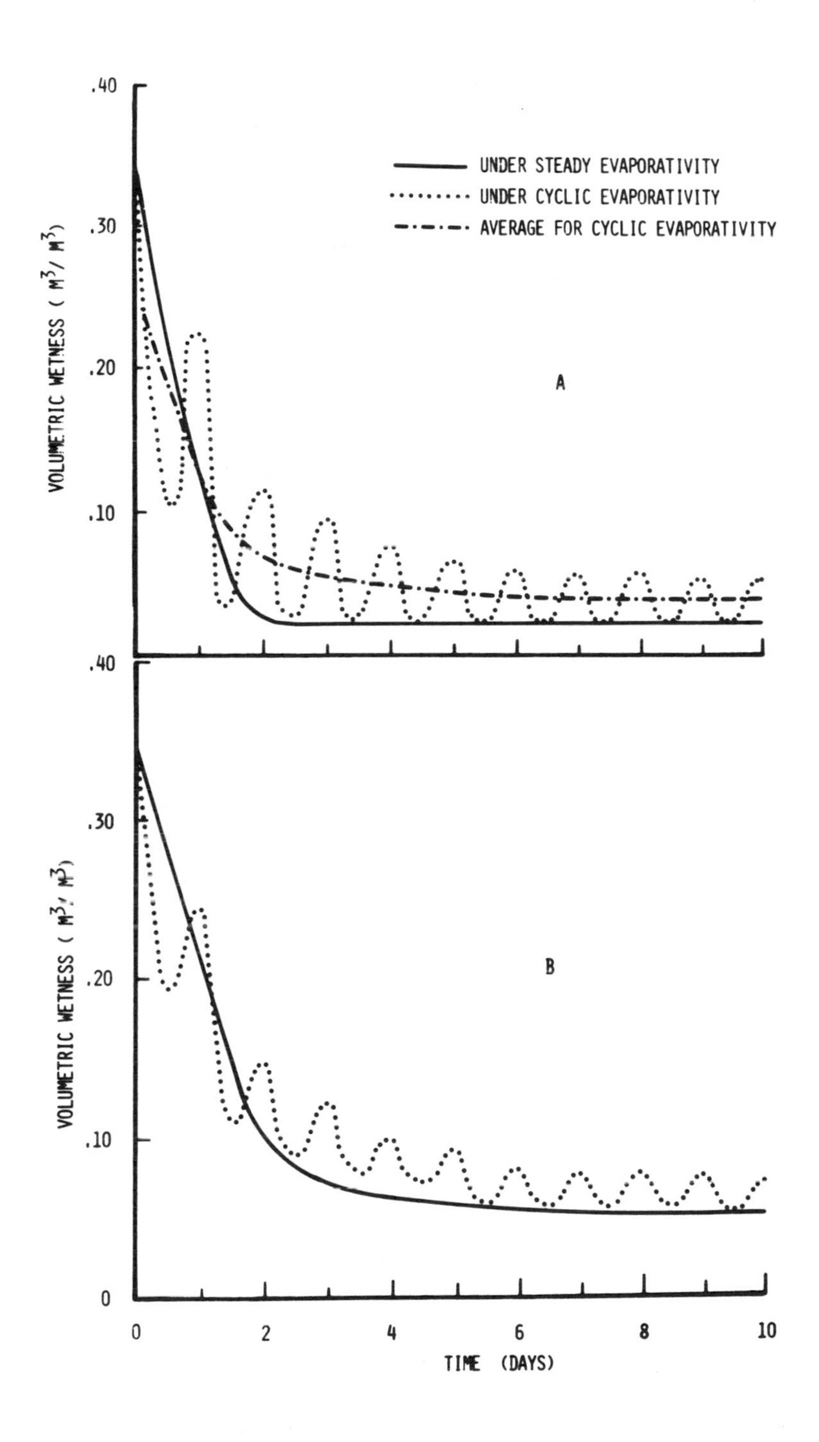

Figure 1.5. Decrease in volumetric wetness of the soil during evaporation: A. At 5 mm depth; B. At 20 mm depth.

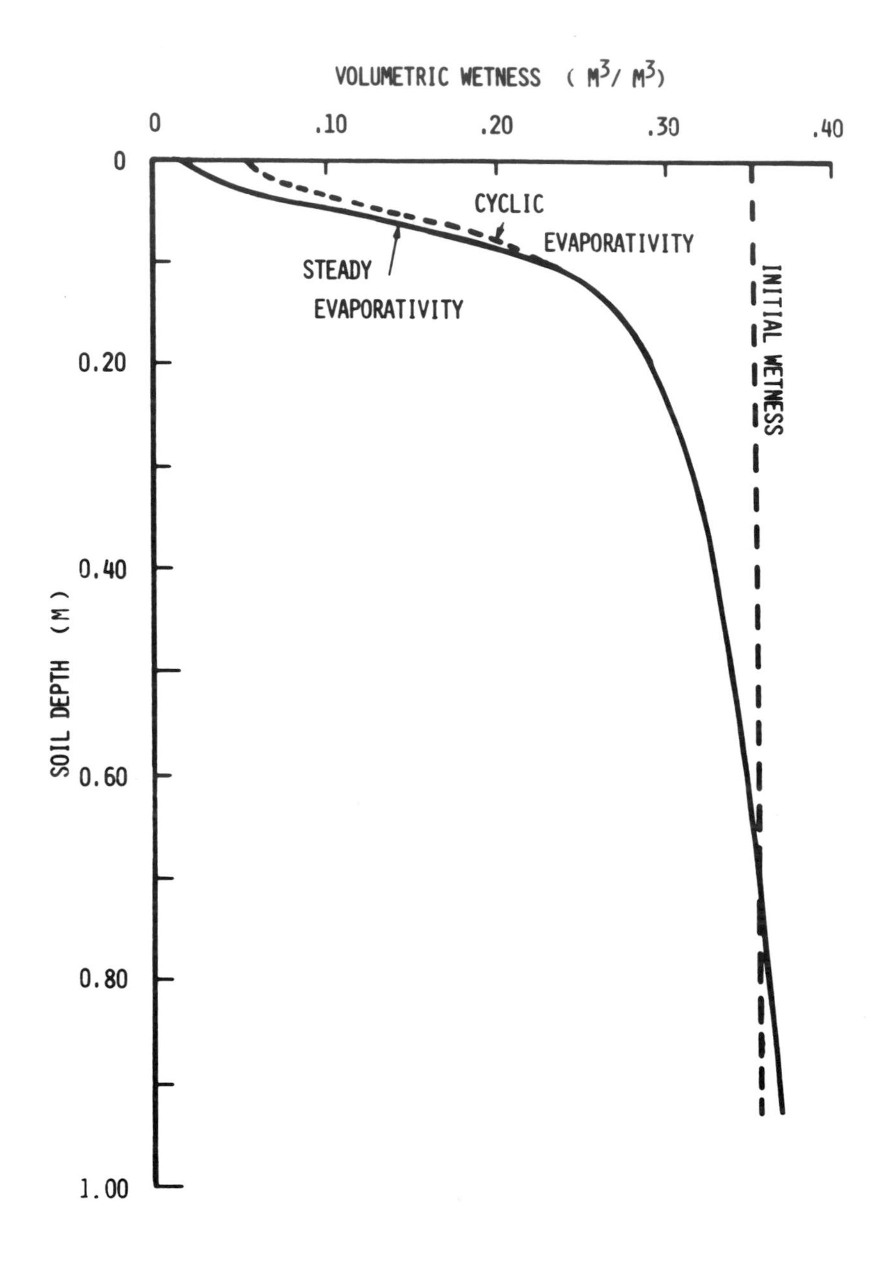

Figure 1.6. Moisture distribution in the soil profile at the end of a 10-day period of simulated evaporation under steady and fluctuating evaporativity.

Discussion

The model presented thus far is obviously a highly simplified representation of the complex processes and interactions which can take place in the field. Specifically, the model described herein is based on isothermal flow theory. As such, it takes no account of temperature gradients and of thermally induced vapour transport. Moreover, since our model assumes uniform soil conditions, it cannot, in its present form, describe cases in which the soil surface zone differs in texture or structure from the underlying soil, or in which the surface is covered by a mulch. Similarly, the present model makes no provision for the simultaneous extraction of soil moisture by plant roots as well as by direct evaporation. Finally, hysteresis of soil moisture characteristics has thus far been ignored.

Notwithstanding these limitations, our model does predict the phenomenon of cyclic drying and resorption of soil moisture in the top layer as it appears under typically fluctuating evaporativity. It further suggests that the classical concept of two or three distinct stages of evaporation is probably meaningless in the context of fluctuating evaporativity.

To ascertain whether the results obtainable by this model are sufficiently realistic for any specific set of conditions, detailed measurements are necessary. Measurements made by the author in Gilat, Israel, indicate a day-night fluctuation of 2 to 3 percent volumetric moisture in the apparently dry surface centimetre of a bare soil, which indeed corresponds to the range of variation predicted by our model after the initial drying phase of, say, the first few days. However, the results obtainable by an isothermal model should be compared with more detailed experimental measurements, and eventually to the predictions obtainable by means of a more comprehensive model based on the simultaneous flow of moisture, heat,and possibly also of solutes, in the soil surface zone, taking hysteresis into account.

F. Modification of Model to Account for the Effect of Hysteresis on Cyclic Evaporation

Definable as the dependence of the equilibrium state of soil moisture (namely, the relation between wetness and suction) upon the direction of the antecedent process (whether sorption or desorption), soil moisture hysteresis was first studied by Haines (1930), and later by Miller and Miller (1956), Youngs (1960), Poulovassilis (1962) and many others. More recently, several investigators (*e.g.*, Rubin 1967; Bresler *et al.* 1969; Vachaud and Thony 1971) have studied the effect of hysteresis on soil water dynamics, particularly during the postinfiltration redistribution phase, in which hysteresis appears to inhibit the downward drainage of water from the infiltration-wetted top layer.

While we still do not have any direct experimental evidence on the role of hysteresis in the diurnal cycle of evaporation, we can conjecture in principle that it might reinforce the tendency toward retardation of evaporation resulting from the diurnal fluctuation of evaporativity. After its strong daytime desiccation, the surface zone of the soil draws moisture from below during the night, so that the top layer is in a process of sorption while the underlying donor layer is in a process of desorption. In principle, the hysteresis phenomenon makes it possible for a sorbing zone of soil to approach potential equilibrium with a desorbing zone of the same soil while the former is at a lower moisture content, and hence at a lower value of hydraulic conductivity. It would seem to follow that the hysteresis effect can contribute to the self-arresting tendency of the evaporation process by causing it to fall below the potential rate earlier than it would if hysteresis were nonexistent.

We therefore modified our evaporation model to test this hypothesis and to indicate the possible extent of the conjectured effect of hysteresis upon the rate and cumulative amount of evaporation over a period of several days.

Description of the Submodel

This section should be considered in conjunction with the program given in Figure 1.2. The INITIAL section of the modified program includes five soil moisture characteristic functions, four more than the original program. The original FUNCTION SUCTB (a table of volumetric wetness versus matric suction) corresponds to the primary desorption curve, henceforth to be designated PDC, as measured for Gilat fine sandy loam from southern Israel. The other four tabulated functions (SUCTW1, SUCTW2, SUCTW3, SUCTW4) represent hypothetical sorption curves of successively greater displacement from PDC, as shown in Fig. 1.7.

Three methods of transition from a desorption to a sorption curve, and vice versa, were tried and their results compared: instantaneous, delayed, and scanning transition.

To provide instantaneous transition, a single statement was added to the DO 200 loop in the DYNAMIC section of the program. Following the original procedure for calculating the matric potential at each layer (MPOT(I)) by interpolation within the SUCTB function, *i.e.*,

```
MPOT(I) = -AFGEN(SUCTB, THETA(I))
```

the following statement was included:

```
IF (NFLUX(I).GE.O.)MPOT(I) = AFGEN(SUCTW1, THETA(I)) where
```

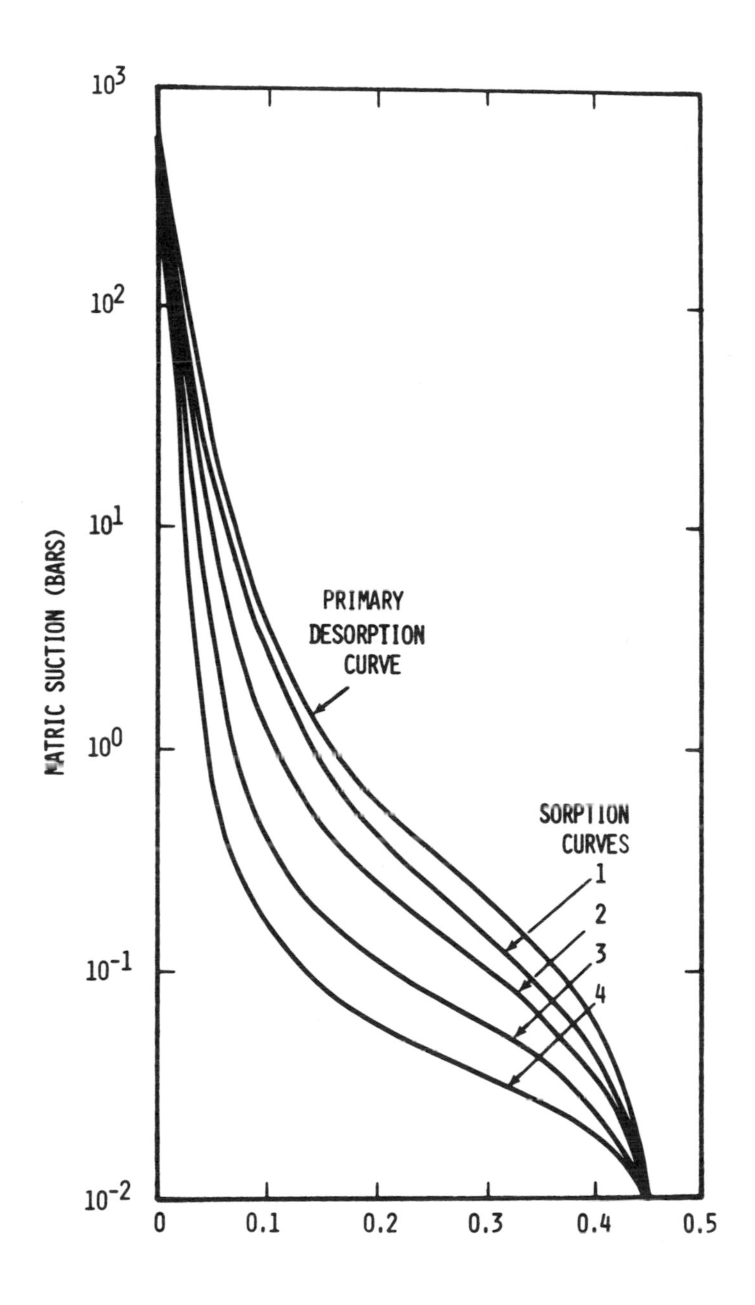

Figure 1.7. The soil moisture characteristic curves used to test the effect of hysteresis on evaporation, including a primary desorption curve and several hypothetical sorption curves.

NFLUX(I), the "net flux," stands for the change of water content of any compartment, indexed (I), during any time increment. If NFLUX(I) is greater than zero, then the soil element in question is in the process of sorbing water, and its characteristic function is immediately shifted from the desorption curve embodied in SUCTB to any one of the sorption curves we may wish to designate, whether SUCTW1, SUCTW2, SUCTW3, or SUCTW4.

To provide delayed transition between the desorption and sorption functions, the following sequence of statements is inserted into the DO 200 loop in lieu of the MPOT(I) statement:

```
          MIMPLS = IMPULS(360.,360.)
          HIMPLS = IMPULS(3600.,3600.)
          IF(MIMPLS.NE.1.)GO TO 888
          DO 777 I=1,NL
      777 TNFLW(I) = TNFLW(I)+NFLUX(I)
      888 CONTINUE
          IF(HIMPLS.NE.1.)GO TO 1001
          DO 999 I=1,NL
          K = 1
          IF(TNFLW(I).GE.0.0)K = 2
      999 TNFLW(I) = 0.
     1001 CONTINUE
     2001 MPOT(I) = -AFGEN(SUCTB, THETA(I))
          GO TO(2001,2002)K
          GO TO 2003
     2002 MPOT(I) = -AFGEN(SUCTW1, THETA(I))
     2003 CONTINUE
```

This procedure in CSMP instructs the computer to read the NFLUX(I) value every 360 seconds and to sum up these values for each elapsed hour (3600 seconds). Only if the *hourly* total of the net flux, TNFLUX(I), is positive, matric potential MPOT(I) is determined from the specified sorption function; otherwise it is determined from the desorption function, SUCTB.

To simulate scanning transition between the desorption and sorption functions, the following subroutine is inserted into the DO 200 loop:

```
          ALT(I) = 0.
          A = AFGEN(SUCTB, THETA(I))
          B = AFGEN(SUCTW1, THETA(I))
          IF(NFLUX(I))500.500.600
      500 IF(ALT(I).EQ.0.)TS(I) = THETA(I)
          ALT(I) = 1.
          X = B+(A-B)*(1.-EXP(ALFA*ABS(THETA(I)-TS(I))))
          GO TO 700
      600 IF(ALT(I).EQ.1.)TS(I) = THETA(I)
          ALT(I) = 0.
```

```
      X = A-(A-B)*(1.  -EXP(ALFA*ABS(THETA(I)-TS(I))))
  700 MPOT(I) = -X
```

Herein, X is the suction during scanning corresponding to the wetness THETA(I), A is the suction value corresponding to the wetness value of a previously desorbing soil layer at the moment it begins to sorb water, B is the suction value of a previously sorbing layer at the moment it begins to desorb, and ALFA is a constant empirically set equal to -0.5. This procedure allows the suction wetness function to loop back and forth on an exponential scanning curve within any interval between two specified primary branches of the hysteretic soil moisture characteristic.

G. Results of a 10-Day Simulation With Hysteresis Taken Into Account.

The results in principle of introducing the hysteretic phenomenon into our simulation of evaporation are summarized in Table 1.1.

The data indicate a systematic reduction of cumulative evaporation with increasing magnitude of the hysteresis range (as indicated by the relative displacement of the four sorption curves from the primary desorption curve, shown in Fig. 1.7). Moreover, the data indicate that the delayed transitions between the sorption and desorption curves consistently yielded lower values of evaporation than the corresponding instantaneous transitions, while the scanning transition gave intermediate values. At its greatest, however, the reduction of evaporation due to hysteresis, as calculated, amounted to about 33 percent of the nonhysteretic evaporation.

The comparison shown in Table 1.1 between the cumulative evaporation where the primary desorption curve (PDC) was used without hysteresis (37.8mm) and the case where sorption curve no. 4 was used without hysteresis (19.6 mm) in effect pertains to two different soils having different pore size distributions. The soil represented by the latter case is more like a coarse-texture soil than like a loam such as is represented by PDC. The data suggest that differences in soil moisture characteristics can strongly influence cumulative evaporation, but this point is incidental to the main thrust of the present paper.

Since the differences among the various assumed hysteretic ranges in terms of total evaporation were rather small (though systematic), only the extreme values were included in the subsequent figures (*i.e.*, the no-hysteresis values of PDC versus the instantaneous and delayed hysteresis values for the reciprocal transition between PDC and the function represented by SUCTW4, with the latter designated SC4).

TABLE 1.1.

Comparison of 10-day cumulative evaporation values (mm) from a bare soil: without hysteresis, and with four assumed ranges of hysteresis calculated on the basis of instantaneous, time-delayed, or scanning transitions between the desorption and sorption characteristic functions.

Characteristic functions:	Instantaneous transition	Delayed transition	Scanning-loop transition
SUCTB-SUCTW1	36.4	35.0	36.0
SUCTB-SUCTW2	34.8	31.3	33.2
SUCTB-SUCTW3	33.1	26.7	32.3
SUCTB-SUCTW4	32.9	25.4	31.3
SUCTB, no hysteresis	37.8		
SUCTW4, no hysteresis	19.6		
Potential evaporation	99.9		

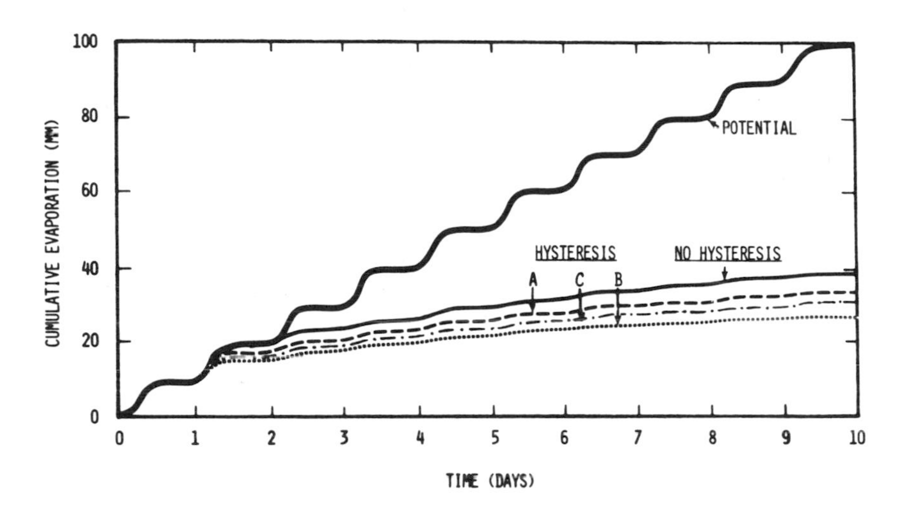

Figure 1.8. Cumulative evaporation computed under diurnally cyclic evaporativity (bold line). Solid line: assuming primary desorption curve (PDC) without hysteresis. Dashed line (A): assuming instantaneous hysteresis between PDC and sorption curve 4 (SC4). Dotted line (B): assuming delayed hysteresis between PDC and SC4. Dot-dash line (C): assuming scanning loops.

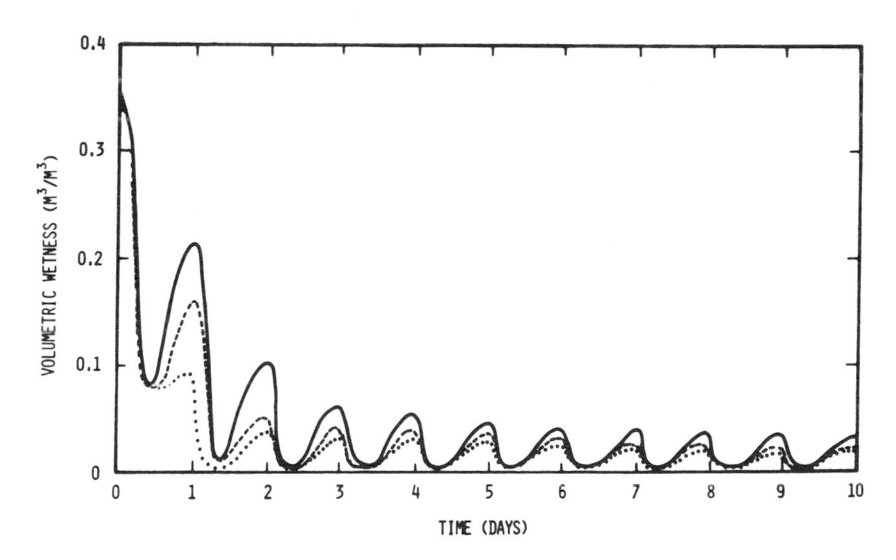

Figure 1.9. Change in volumetric wetness of soil surface zone (5 mm depth) during diurnally cyclic evaporation. Solid line: assuming primary desorption curve (PDC) without hysteresis. Dashed line: assuming instantaneous hysteresis between PDC and sorption curve 4 (SC4); dotted line: assuming delayed hysteresis between PDC and SC4. The pattern for scanning hysteresis was intermediate between those for instantaneous and delayed hysteresis.

The time-course of cumulative evaporation for each of these characteristic functions is illustrated in Fig. 1.8.

The probable reason for these differences is illustrated in Fig. 1.9, which gives the continuous variation of the surface-zone moisture content during the 10 successive diurnal cycles of the simulation period. All three soil conditions illustrated are seen to behave similarly only during the first half day of monotonic drying. Already in the afternoon of that first day, however, as the evaporational demand falls below its noontime peak (eventually to become zero during the night) differences begin to occur in the rate of rewetting of the surface zone. The resorption process is markedly inhibited in the hysteretic soil as against the nonhysteretic soil, and more so where the transition from the sorption to the desorption characteristic functions was delayed rather than instantaneous. This effect is repeated during each afternoon and nighttime resorption phase, and is still discernible during the daily desorption phases.

Finally, Fig. 1.10 gives the distribution of soil moisture at the end of the 10-day period of simulated evaporation, as affected by hysteresis. The hysteretic profiles indicated lower water contents at the top 10 cm or so, and appreciably higher water contents throughout the lower half-metre or so, than in the case of the nonhysteretic profile. This is an additional reflection of the role of hysteresis in reducing the tendency of the uppermost soil layer to resorb water by suction from the underlying moist layers during periods of low or zero evaporation.

Discussion

The present simulation of evaporation from a uniform soil, while more realistic than the original hydraulic model upon which it was based, still does not take into account such factors as thermal gradients, vapour transfer, and the possible presence of a surface mulch. While such factors have been included in recent simulation models (*e.g.*, van Bavel and Hillel 1975; Hillel *et al.* 1975), they were not essential to the limited objective of this study, which was to determine the specific role of hysteresis during evaporation. Another principal aspect of our simulation is the assumption that hydraulic conductivity is a unique and single-valued function of soil wetness, regardless of the hysteretic nature of the suction function. Although this assumption is widely accepted as a working hypothesis at present, it may justify closer scrutiny before it can be taken for granted in general.

The procedures we used for the transition between the sorption and desorption characteristic functions can also be faulted for being arbitrary. Here again we are faced with insufficient knowledge of the transition characteristics of real soils. Although scanning loops have been studied rather extensively,

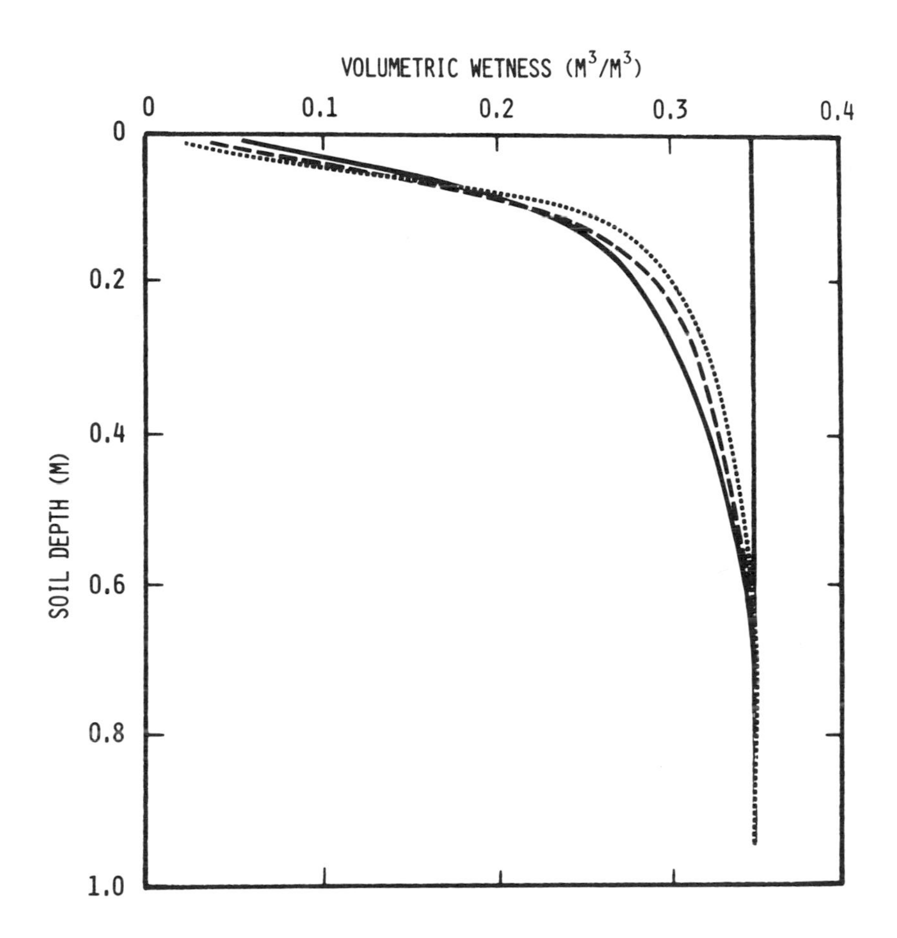

Figure 1.10. Moisture distribution in the soil profile at the end of a 10-day period of simulated evaporation, under diurnally cyclic evaporativity. Solid line: assuming primary desorption curve (PDC) without hysteresis. Dashed line: assuming instantaneous hysteresis between PDC and sorption curve 4 (SC4); dotted line: assuming delayed hysteresis between PDC and SC4.

there are still conflicting opinions on the shape and uniqueness of these curves (*e.g.*, Davidson *et al.* 1966; Vachaud and Thony 1971). In modeling hysteretic effects, Whisler and Klute (1965) used arbitrary scanning curves, Bresler *et al.* (1969) assumed straight-line scanning with a slope dependent on the transition point, whereas Staple (1969) determined intermediate scanning curves by interpolation between the main hysteresis branches of the soil moisture characteristic.

In our case, the results obtained with three methods of transition, though arbitrary, appear to be sufficient to confirm in principle that the hypothesis regarding the inhibiting effect of hysteresis on cyclic evaporation is essentially valid. This finding, in turn, suggests the development of a method to reduce evaporation artificially by a treatment designed to cause the rapid desiccation of the soil surface. The possible magnitude of the effect, and the extent to which it can be induced and controlled in practice, remain to be tested in actual field conditions.

II. NON-ISOTHERMAL EVAPORATION OF SOIL WATER, INCLUDING THE EFFECT OF SURFACE REFLECTIVITY

A. Description of the Problem

In the model of isothermal evaporation given in the previous chapter, potential evaporation (evaporativity) was introduced as an external forcing function. No account was taken of the surface radiation balance or of the soil's energy balance, including such processes as absorption and conduction of heat by the soil and exchange of heat between the soil and the atmosphere. Thus, the thermally-induced transport of vapour in the soil surface zone was completely ignored.

During non-isothermal evaporation, coupled flow takes place between vapour and liquid water. Vapour can diffuse through the continuous air-filled pores. However, vapour transport has been found to exceed the rate predictable on the basis of diffusion alone. According to Philip and de Vries (1957), vapour movement apparently occurs by a complex sequential process of evaporation, short-range diffusion, condensation in capillary pockets of liquid, short-range liquid flow, re-evaporation, etc. Liquid flow is probably the dominant mechanism of water movement in wet, nearly isothermal soil, whereas coupled flow probably takes place at intermediate moisture contents and where temperature gradients are appreciable; finally, vapour diffusion is likely to be the dominant mechanism of water transport in the low moisture range (such as in a desiccated surface-layer).

A rather comprehensive simulation model of non-isothermal evaporation, based on the use of actual weather data (radiation, air temperature and humidity, and wind speed) as inputs, and allowing the calculation of evaporative demand rather than its imposition as a forcing function, has recently been published by van Bavel and Hillel (1975, 1977). In addition to the storage and disposition of water in the soil profile, this model can predict partitioning of radiant energy into sensible and latent heat and the resulting pattern of soil temperature. This chapter is based largely on that model.

Knowledge of soil moisture and temperature can be important in field practice, particularly insofar as these can be affected by controllable factors such as surface albedo, roughness, and the presence of a mulch. The need for a simulation approach arises because of the highly complex and dynamic nature of so many variables and interactions which together comprise the system of interest. It seems well-nigh impossible to obtain analytical solutions of the governing equations of the system without introducing excessively restrictive, oversimplified, or unrealistic assumptions.

B. Governing Equations

We begin with the radiation balance of a bare soil surface, which can be written as follows:

$$J_n = (s_s + s_a)(1 - a) + \ell_i - \ell_e \quad (2.1)$$

Herein, J_n is the net radiation, s_s the incoming flux of short-wave radiation from the sun; s_a the incoming short-wave radiation flux from the atmosphere (sky); ℓ_i the incoming long-wave radiation flux from the sky; ℓ_e the long-wave radiation emitted by the soil; and a the reflectivity coefficient, called *albedo*.

The global short-wave radiation flux $(s_s + s_a)$ is often obtainable from meteorological measurements; ℓ_i is proportional to the effective air temperature raised to the fourth power, and can be estimated according to the Brunt formula (Sellers 1965), provided the air temperature at some specified height (T_a) and humidity (H_a) are known:

$$\ell_i = \sigma T_a{}^4 [0.605+0.039(1.41H_a)^{\frac{1}{2}}] \quad (2.2)$$

where σ is the Stefan-Boltzmann constant; ℓ_e is similarly proportional to the fourth power of the soil surface temperature (T_s):

$$\ell_e = \varepsilon \sigma T_s{}^4 \quad (2.3)$$

Here ε is the emissivity, a function of soil surface wetness. Finally, the albedo a is known to depend on the natural colour and wetness of the soil surface. It generally ranges between 10-35%, usually increasing as the soil dries; however, it can be modified by various surface treatments.

The net radiation received by the soil surface is transformed into heat which warms the soil and the air and vaporizes water. We can thus write the surface energy balance as follows:

$$J_n = S + A + LE \quad (2.4)$$

where S is the soil heat flux (the rate at which heat is taken up by the soil); A is the sensible heat flux going into the air overlying the soil surface; and LE is the evaporative heat flux (a product of the evaporation rate E and the latent heat per unit quantity of water evaporated, L).

The total surface energy balance [equations (2.1) and (2.4)] is therefore:

$$(s_s + s_a)(1 - a) + \ell_i - \ell_e - S - A - LE = 0 \qquad (2.5)$$

Conventionally, all components of the energy balance are taken as positive if directed toward the surface, and negative otherwise.

The simulation of soil heat flow was described by Wierenga and de Wit (1970). The governing equations were recently summarized by de Vries (1975). The differential equation for heat conduction in soil is given by *Fourrier's law*,

$$q_h = -\lambda \nabla T \qquad (2.6)$$

which, in one-dimensional form, is written as

$$q_h = -\lambda \frac{dT}{dx} \qquad (2.7)$$

In these equations, q_h is thermal flux, λ thermal conductivity, and dT/dx the temperature gradient. Another form of the same equation is

$$q_h = -\frac{D_h}{C} \frac{dT}{dx} \qquad (2.8)$$

where D_h is thermal diffusivity and C volumetric heat capacity. The energy conservation (continuity) equation is:

$$C \frac{\partial T}{\partial t} = -\nabla q_h \qquad (2.9)$$

where t is time. The combined equation is thus:

$$C \frac{\partial T}{\partial t} = \nabla(\lambda \nabla T) \qquad (2.10)$$

For one-dimensional flow we have:

$$C \frac{\partial T}{\partial t} = \frac{\partial}{\partial x}\left(\lambda \frac{\partial T}{\partial x}\right) \qquad (2.11)$$

Both volumetric heat capacity and thermal conductivity depend upon soil particle composition, bulk density, and wetness:

$$C = \Sigma f_{si} C_{si} + f_w C_w + f_a C_a \qquad (2.12)$$

Here, f denotes the volumetric fraction of each phase: air (subscripted a), water (w), and solids (including a number of solid constituents, subscripted i). The C value for water, air, and each solid constituent is the product of the particular density and specific heat.

If we disregard the air phase and distinguish within the solid phase only between mineral matter (to be subscripted m) and organic matter (o), we obtain:

$$C = f_{sm}C_{sm} + f_{so}C_{so} + f_wC_w \qquad (2.13)$$

with $f_{sm} + f_{so} + f_w = 1-f_a$. (The total porosity $f_p = f_a+f_w$).

Thermal properties and densities of various soil constituents were given by de Vries (1975). Thermal conductivity of soil can be calculated from the volumetric fraction of water (wetness) as follows:

$$\lambda = \frac{f_w\lambda_w + \Sigma k_i f_i \lambda_i + k_a f_a \lambda_a}{f_w + \Sigma k_i f_i + k_a f_a} \qquad (2.14)$$

wherein λ_w, λ_a and λ_i are the specific thermal conductivities of each of the soil constituents (water, air, and each of the solid components, respectively). The factors k_i represent ratios between the space averages of the temperature gradients in each solid constituent type i and the space average of temperature gradient in the water phase. The k_i values depend on grain shape as well as composition. The k_a factors represent the ratios for thermal gradients in the air and water phases in the soil.

The influence of latent heat transfer by water vapour in the air-filled pores is proportional to the temperature gradient in these pores. It can be taken into account by adding to the thermal conductivity of air an apparent conductivity due to evaporation, transport, and condensation of water vapour. This value is strongly temperature dependent. The flux of sensible heat in the soil associated with liquid water movement is considered negligible in our model.

Returning now to the energy balance equation (2.4), we can express the transport of sensible heat from the soil to the boundary layer of the atmosphere as follows:

$$A = -c_p\rho_a\kappa_a \frac{dT}{dz} \qquad (2.15)$$

where c_p is the specific heat capacity of the air at constant pressure, ρ_a the density of air, κ_a the turbulent transfer coefficient for heat, and dT/dz the temperature gradient with height z above the soil surface. The reciprocal of κ_a is the aerodynamic resistance r_a, which can be calculated from the known windspeed and surface roughness. In case the temperature profile of the air is not adiabatically neutral, a stability correction can be applied (Monteith 1963; Szeicz *et al.* 1973).

The rate of latent heat transfer by water vapour from soil to atmosphere, LE, is similarly proportional to the product of the vapour pressure gradient and the appropriate turbulent transfer coefficient for vapour.

To express these equations of heat and vapour transport as

finite difference equations to calculate soil heat and moisture dynamics, we should know the surface temperature and vapour pressure. In our model, we use the implicit-loop procedure of CSMP to calculate the former by a process of iteration. The latter can then be calculated (Staple 1974) from the water potential equivalent of the surface soil, ϕ_s, according to:

$$\phi_s = RT_s ln \frac{p_s}{p_o} \qquad (2.16)$$

where p_s is the vapour pressure of soil water and p_o is the saturation vapour pressure at the same temperature, T_s. The ratio p_s/p_o is the relative humidity. This equation can be rewritten as:

$$p_s = p_o \exp(\phi_s/RT_s) \qquad (2.17)$$

In these equations, R is, of course, the universal gas constant.

C. Description of the Computer Model

Following the work of van Bavel and Hillel (1975, 1977), the computer model used for the non-isothermal evaporation of soil moisture is shown in Figure 2.1. The hydraulic aspect of the model, as it pertains to the conduction of water in the profile, is essentially identical with the isothermal model of the previous chapter. The present model differs, however, in its inclusion of meteorological inputs, parameters, and equations pertaining to the calculation of the soil energy balance and thermal regime. Note that KOND is soil thermal conductivity, FLOW soil heat flux, TEMP soil temperature, VHCAP volumetric heat capacity (ITEMP and IVHCAP being the corresponding initial values), VOLH heat content, NFLOW net flow of heat — all subscripted as they refer to each compartment in the profile.

INITIAL Section

This section begins with a list of the parameters, the first of which is the albedo, ALB, for which repeated runs are ordered at values of 0.05 and 0.50. In a separate simulation experiment the albedo was taken to be a function of the wetness of soil in the top compartment, varying linearly from 0.10 at volumetric wetness values of 25% (or above) to a maximal value of 0.35 for soil wetness values of 10% (or below). The other parameters are the soil's porosity, PORSTY, assumed to remain constant regardless of soil wetness; acceleration of gravity, GRAV; Stefan-Boltzmann constant, SIGMA; surface roughness length, ZO; latent heat of evaporation, L; volumetric heat capacity of air, VHCAPA, of soil solids, VHCAPS, and of water, VHCAPW; thermal conductivity of air, KONDA, of soil solids, KONDS, of liquid water, KONDW, and of water vapour, KONDV. Note that S.I. units (metre-kilogram-second-ampere) were used throughout this program, as well as

Figure 2.1. CSMP listing for the dynamic simulation of soil moisture and energy interrelations and the effect of albedo on evaporation and soil temperature.

```
TITLE   SIMULATION OF WATER AND ENERGY TRANSPORT IN SOIL PROFILE
*              DURING EVAPORATION:  TEST OF ALBEDO EFFECT

*              UNITS
*    C  = DEGREES CELSIUS
*    J  = JOULES
*    K  = DEGREES KELVIN
*    KG = KILOGRAMS
*    M  = METERS
*    S  = SECONDS
*    W  = WATTS

*                          GLOSSARY OF SYMBOLS

* A      = SENSIBLE HEAR FLUX IN AIR, W/(M**2)
* ALB    = ALBEDO, FRACTION
* AVCOND = AVERAGE HYDRAULIC CONDUCTIVITY, M/S
* AVKOND = AVERAGE THERMAL CONDUCTIVITY, W/(M*S)
* COND   = HYDRAULIC CONDUCTIVITY, M/S
* CUMDRN = CUMULATIVE DRAINAGE, M
* CUMEV  = CUMULATIVE EVAPORATION, M
* DEPTH  = DEPTH OF CENTER OF COMPARTMENT, M
* DIST   = DISTANCE BETWEEN ADJACENT COMPARTMENT CENTERS, M
* DNUM   = DAY NUMBER
* DP     = DEWPOINT TEMPERATURE OF AIR, C
* EM     = EMITTANCE, FRACTION
* EV     = EVAPORATION RATE, M/S
* EV2    = EV
* FLOW   = SOIL HEAT FLOW RATE, W/(M**2)
* FLUX   = WATER FLOW RATE, M/S
* FTS    = FINAL VALUE OF TS
* GR     = GLOBAL IRRADIANCE, W/(M**2)
* GRAV   = ACCELERATION OF GRAVITY, M/(S**2)
* HA     = HUMIDITY OF AIR, KG/(M**3)
* HO     = HUMIDITY AT SOIL SURFACE, KG/(M**3)
* HO2    = HO
* HOS    = SATURATION HUMIDITY AT SURFACE TEMPERATURE, KG/(M**3)
* HOS2   = HOS
* HPOT   = HYDRAULIC POTENTIAL HEAD, M
* ITEMP  = INITIAL TEMPERATURE, C
* ITHETA = INITIAL SOIL WETNESS, VOLUME FRACTION
* IVHCAP = INITIAL VOLUMETRIC HEAT CAPACITY, J/(C*M**3)
* IVOLH  = INITIAL HEAT CONTENT OF COMPARTMENT, J/(M**2)
* IVOLW  = INITIAL VOLUME OF WATER IN COMPARTMENT, M
* KO     = KONDV
```

```
* KOND   = THERMAL CONDUCTIVITY OF SOIL, W/(M*S)
* KONDA  = THERMAL CONDUCTIVITY OF AIR, W/(M*S)
* KONDS  = THERMAL CONDUCTIVITY OF SOIL SOLIDS, W/(M*S)
* KONDV  = THERMAL CONDUCTIVITY OF WATER VAPOUR, W/(M*S)
* KONDW  = THERMAL CONDUCTIVITY OF WATER LIQUID, W/(M*S)
* LE     = LATENT HEAT FLUX IN AIR, W/(M**2)
* LH     = LATENT HEAT OF VAPORIZATION, J/KG
* MPOT   = MATRIC POTENTIAL, M
* NFLOW  = NET FLOW OF HEAT INTO COMPARTMENT, W/(M**2)
* NFLUX  = NET FLOW OF WATER INTO COMPARTMENT, M/S
* NL     = NUMBER OF COMPARTMENTS
* NR     = NET RADIATION, W/(M**2)
* PORSTY = POROSITY OF SOIL, VOLUME FRACTION
* PP     = PPOT
* PPOT   = PRESSURE POTENTIAL, M
* RA     = RESISTANCE OF AIR BOUNDARY LAYER, S/M
* RAC    = RESISTANCE OF AIR BOUNDARY, CORRECTED FOR STABILITY, S/M
* RAC2   = RAC
* RHO    = RELATIVE HUMIDITY AT SURFACE, FRACTION
* RI     = RICHARDSON'S NUMBER
* RI2    = RI
* S      = SENSIBLE HEAT FLUX INTO SOIL, W/(M**2)
* SA     = WIND SPEED, M/S
* SH     = VOLUMETRIC HEAT CAPACITY OF AIR, J/(C*M**3)
* SIGMA  = STEFAN-BOLTZMANN CONSTANT, W/(M**2*K**4)
* SKL    = SKY LONGWAVE IRRADIANCE, W/(M**2)
* STAB   = INDEX OF AIR STABILITY, FRACTION
* ST     = STIME
* STIME  = STANDARD TIME, HOURS
* T1     = VOLUMETRIC WETNESS OF FIRST COMPARTMENT, FRACTION
* TA     = TEMPERATURE OF AIR, C
* TCOM   = THICKNESS OF COMPARTMENT, M
* TE     = TEMP
* TEMP   = SOIL TEMPERATURE, C
* TH     = THETA
* THETA  = SOIL WETNESS, VOLUME FRACTION
* TS     = TEMPERATURE OF SOIL SURFACE, C
* VHCAP  = VOLUMETRIC HEAT CAPACITY, J/(C*M**3)
* VHCAPS = VOLUMETRIC HEAT CAPACITY OF SOLIDS, J/(C*M**3)
* VCHAPW = VOLUMETRIC HEAT CAPACITY OF WATER, J/(C*M**3)
* VOLH   = HEAT CONTENT OF COMPARTMENT, J/(M**2)
* VOLW   = VOLUME OF WATER IN COMPARTMENT, M
* ZO     = SURFACE ROUGHNESS PARAMETER, M
```

```
STORAGE        TCOM(25),ITHETA(25),DEPTH(25),COND(25),AVCOND(25),FLUX(25)
STORAGE        MPOT(25),HPOT(25),DIST(25),ITEMP(25),KOND(25),AVKOND(25)
STORAGE        FLOW(25),VHCAP(25),IVHCAP(25)
/      DIMENSION  THETA(25),VOLW(25),IVOLW(25),NFLUX(25)
/      DIMENSION  TEMP(25),VOLH(25),IVOLH(25),NFLOW(25)
/      EQUIVALENCE  (VOLW1,VOLW(1)),(IVOLW1,IVOLW(1)),(NFLUX1,NFLUX(1))
/      EQUIVALENCE  (VOLH1,VOLH(1)),(IVOLH1,IVOLH(1)),(NFLOW1,NFLOW(1))
FIXED          I,NL,NLL

INITIAL

NOSORT
PARAMETER ALB=(.05,.50)
PARAMETER PORSTY=.45,GRAV=9.81,SIGMA=6.57E-8,ZO=.01,LH=2.442E9
PARAMETER SH=1.1E3,VHCAPS=1.925E6,VHCAPW=4.186E6,KONDS=4.2
PARAMETER KONDW=.57,KONDA=.025
               NL=18
               NLL=NL+1
TABLE TCOM(1-18)=2*.01,2*.02,6*.04,9*.10
TABLE ITHETA(1-18)=18*.45
               DEPTH(1)=.5*TCOM(1)
               DIST(1)=DEPTH(1)
       DO 110 I=2,NL
               DIST(I)=.5*(TCOM(I-1)+TCOM(I))
           110 DEPTH(I)=DEPTH(I-1)+.5*(TCOM(I-1)+TCOM(I))
       DO 120 I=1,NL
               NFLUX(I)=0.
               NFLOW(I)=0.
               ITEMP(I)=25.
               IVHCAP(I)=2.12E6
               IVOLH(I)=ITEMP(I)*TCOM(I)*IVHCAP(I)
           120 IVOLW(I)=ITHETA(I)*TCOM(I)
FUNCTION SUCTB=(.001,100000.),(.005,10000.),(.01,3500.),(.025,1000.),...
       (.05,200.),(.1,40.),(.15,10.),(.2,6.),(.25,3.5),(.3,2.2),       ...
       (.35,1.4),(.4,.56),(.45,0.),(1.,-100.)
FUNCTION CONDTB=(.001,.4E-15),(.01,.4E-13),(.05,.2E-12),(.1,.14E-11),...
       (.15,.8E-11),(.2,.5E-10),(.25,.3E-9),(.3,.2E-8),(.35,.12E-7),  ...
       (.4,.8E-7),(.45,.5E-6),(1.,.5E-6)
FUNCTION ATEMTB=(0.,23.),(1.,22.5),(2.,22.),(3.,21.5),(4.,21.),       ...
       (5.,20.5),(6.,20.),(7.,20.5),(8.,21.5),(9.,22.5),(10.,24.),    ...
       (11.,25.5),(12.,26.5),(13.,28.),(14.,29.),(15.,29.5),          ...
       (16.,29.6),(17.,29.5),(18.,29.),(19.,28.),(20.,26.5),          ...
       (21.,25.),(22.,24.),(23.,23.5),(24.,23.)
FUNCTION DEWPTB=(0.,14.5),(3.,15.),(6.,15.),(9.,16.),(12.,15.5),      ...
       (15.,15.),(18.,14.),(21.,14.5),(24.,14.5)
FUNCTION WINDTB=(0.,4.),(3.,3.5),(9.,4.5),(12.,5.),(15.,5.),(15.,5.),...
       (18.,5.),(21.,4.),(24.,4.)
FUNCTION RADTB=(0.,0.),(5.,0.),(6.,50.),(7.,230.),(8.,450.),          ...
       (9.,650.),(10.,800.),(11.,900.),(12.,960.),(13.,950.),         ...
       (14.,900.),(15.,750.),(16.,400.),(17.,200.),(18.,50.),         ...
       (19.,0.),(24.,0.)
FUNCTION VAPKTB=(0.,.025),(10.,.04),(20.,.08),(30.,.125),             ...
       (40.,.25),(50.,.40),(60.,.65),(70.,1.2)

DYNAMIC

NOSORT
               STIME=TIME/3600.
               STIME=AMOD(STIME,24.)
               DNUM=TIME/86400.
       DO 200 I=1,NL
               THETA(I)=VOLW(I)/TCOM(I)
               COND(I)=AFGEN(CONDTB,THETA(I))
               MPOT(I)=AFGEN(SUCTB,THETA(I))
               HPOT(I)=MPOT(I)-DEPTH(I)
               VHCAP(I)=VHCAPW*THETA(I)+(1.-PORSTY)*VHCAPS
               TEMP(I)=VOLH(I)/(VHCAP(I)*TCOM(I))
               KONDV=AFGEN(VAPKTB,TEMP(I))
               KOND(I)=((1.-PORSTY)*KONDS*.4+THETA(I)*KONDW+           ...
                  (PORSTY-THETA(I))*1.4*(KONDA+KONDV))/                ...
                  (1.-PORSTY)*.4+THETA(I)+(PORSTY-THETA(I))*1.4
```

```
     200 CONTINUE
  DO 210 I=2,NL
         AVCOND(I)=(COND(I-1)*TCOM(I-1)+COND(I)*TCOM(I))/(TCOM(I-1)+TCOM(I))
     210 AVKOND(I)=(TCOM(I-1)+TCOM(I))/(TCOM(I-1)/KOND(I-1)+TCOM(I)/KOND(I))
         FLUX(NLL)=COND(NL)
         FLOW(NLL)=0.
  DO 220 I=2,NL
         FLUX(I)=(HPOT(I-1)-HPOT(I))*AVCOND(I)/DIST(I)
     220 FLOW(I)=(TEMP(I-1)-TEMP(I))*AVKOND(I)/DIST(I)
         GR=AFGEN(RADTB,STIME)
         T1=THETA(1)
         EM=.90+T1*.08/.45
         SA=AFGEN(WINDTB,STIME)
         RA=(ALOG(2./ZO)**2.)/(.16*SA)
         DP=AFGEN(DEWPTB,STIME)
         HA=1.323*EXP(17.27*DPTC/(237.3+DPTC))/(273.16+DP)
         TA=AFGEN(ATEMTB,STIME)
         SKL=(SIGMA*(TA+273.16)**4.)*(.605+.039*SQRT(1410.*HA)
*
         TS=IMPL(TA,.01,FTS)
         RI=AMIN1(.08,(GRAV*(2.-ZO)*(TA-TS)/((TA+273.16)*SA**2.)))
         RAC=FA/(1.-10.*RI)
         A=(TS-TA)*SH/RAC
         HOS=1.323*EXP(17.27*TS/(237.3*TS))/(273.16+TS)
         HO=HOS*EXP(MPOT(1)/(46.97*(TS+273.16)))
         EV=(HO-HA)/(RAC*1000.)
         S=GR*(1.-ALB)+SKL-EM*SIGMA*(TS+273.16)**4.-A-LH*EV
         FTS=TEMP(1)+S*DEPTH(1)/KOND(1)
*
         FLOW(1)=(TS-TEMP(1))*KOND(1)/DIST(1)
         HOS2=1.323*EXP(17.27*TS/(237.3+TS))/(273.16+TS)
         HO2=HOS2+EXP(MPOT(1)/(46.97*(TS+273.16)))
         RHO=HO2/HOS2
         RI2=AMIN1(.08,(GRAV*(2.-ZO)*(TA-TS)/((TA+273.16)*SA**2.)))
         RAC2=RA/(1.-10.*RI2)
         STAB=1./(1.-10.*RI2)
         EV2=(HO2-HA)/(RAC2*1000.)
         A2=(TS-TA)*SH/RAC2
         LE=LH*EV2
         NR=FLOW(1)+(TS-TA)*SH/RAC2+LH*EV2
         FLUX(1)=-EV2
  DO 320 I=1,NL
         NFLOW(I)=FLOW(I)-FLOW(I+1)
     320 NFLUX(I)=FLUX(I)-FLUX(I+1)
         VOLH1=INTGRL(IVOLH1,NFLOW1,18)
         VOLW1=INTGRL(IVOLW1,NFLUX1,18)
         S2=FLOW(1)
         CUMS=INTGRL(0.,S2)
         FLXNLL=FLUX(NLL)
         CUMEV=INTGRL(0.,EV2)
         CUMNR=INTGRL(0.,NR)
         CUMA=INTGRL(0.,A2)
         CUMDRN=INTGRL(0.,FLXNLL)

TERMINAL

     TIMER  FINTIM=864000., OUTDEL=7200.,DELMIN=.1E-5
     PRTPLT (optional)
     METHOD RKS
     END
     STOP
```

all other programs in this monograph.

The INITIAL segment of this program continues with specifications of the number of layers, NL, and a table of their thicknesses, TCOM, providing a total profile depth of 1.2m; initial wetness, ITHETA; and calculations of depth for each compartment and flow segment lengths between adjacent compartments. The initial temperatures and volumetric heat capacities of all compartments are assigned values, whereas net heat and water fluxes are initiated at zero. The initial heat and water contents of all compartments are then calculated on the basis of the particular compartment thicknesses.

The following seven tabulated functions follow:

(1) SUCTB: Volumetric wetness (fraction) versus matric suction head (metres), pertaining to the moisture desorption characteristic curve for Gilat fine sandy loam of the Northern Negev of Israel.

(2) CONDTB: Volumetric wetness versus hydraulic conductivity (metres/second) for the same soil.

(3) ATEMTB: Standard time (hours) versus air temperature (°C) at standard height.

(4) DEWPTB: Standard time versus dew point temperature of the air (°C).

(5) WINDTB: Standard time versus wind speed (metres/second) at standard height.

(6) VAPKTB: Soil temperature versus the effective thermal conductivity of water vapour.

DYNAMIC Section

The system variable, TIME, is cast into standard time units in hours, STIME, and successive days are made to repeat the same meteorological input variables by means of the AMOD function of CSMP. DNUM is the ordinal number of successive days during the simulation run.

The DO 200 loop calculates, for each compartment and time step, the values of volumetric wetness THETA, hydraulic conductivity COND, matric potential head MPOT, hydraulic potential head HPOT, volumetric heat capacity VHCAP, temperature TEMP, effective thermal conductivity due to vapour movement KONDV, and overall thermal conductivity KOND. The latter calculations are carried out according to the equations presented in the preceding section of this chapter.

The DO 210 loop calculates the average thermal conductivity AVKOND for each inter-compartmental flow segment. The bottom boundary condition is then set for gravity drainage (FLX(NLL) = COND(NL)), but no heat flow (FLOW(NLL) = 0). The DO 220 loop then calculates the internal fluxes of water and heat in the profile from Darcy's and Fourrier's laws.

To set the stage for the calculation of the energy balance and the fluxes of heat and water vapour through the soil surface, we determine and update the following dynamic variables: global irradiance GR, volumetric wetness of the top compartment T1, surface emittance EM, wind speed SA, turbulent resistance coefficient for the atmospheric boundary layer RA, dewpoint temperature of the air DP, air humidity HA, air temperature TA, and sky long-wave irradiance SKL.

The implicit loop procedure for the iterative calculation of the soil surface temperature TS follows. The subprogram consists of nine statements, the first of which is

```
TS = IMPL(TA, 0.01, FTS)                                    (2.18)
```

This establishes the air temperature, TA, as an initial guess for TS, with 0.01 as the acceptable convergence error for the value of TS, and FTS as the dummy name used within the iteration for TS, to be checked against the guessed value of TS each time.

Other variables calculated include Richardson's number RI, the turbulent resistance coefficient for the atmospheric boundary layer RAC (corrected for adiabatic instability), the sensible heat transfer from soil to air A, the saturation air humidity at the soil surface HO (a combined function of the matric potential MPOT of the profile's top compartment and the surface temperature TS), the evaporation rate EV (equal to the rate of vapour transport from soil surface to the air), and heat flux into the soil S (obtained from the overall energy balance).

The last statement of the implicit loop calculates FTS by rearrangement of the equation for heat flow between the surface and the centre of the topmost compartment:

```
S = KOND(1)*(FTS-TEM(1))/DEPTH(1)                           (2.19)
```

Emerging from the implicit loop with an acceptable value of surface temperature, we can now calculate the "true" values of the heat flux into the soil FLOW(1); the saturation humidity at the soil surface HOS2, as well as the actual humidity HO2 and relative humidity RHO at that level; Richardson's number RI2; the atmospheric boundary-layer resistance to both heat and vapour transfer RAC2; air stability index STAB; evaporation rate EV2, sensible heat flux into the air A2, and latent heat flux into the air; net radiation NR; and the flux of water through the soil surface FLUX(1), which is numerically equal to the evaporation rate but of opposite sign.

The DO 320 loop then computes the net increments of heat and water (NFLOW and NFLUX, respectively) for each compartment and time step. The total content of heat and water in each compartment is integrated by means of the INTGRL statement, using METHOD RKS (Runge-Kutta). Similar integral statements follow for the cumulative values of evaporation, net radiation, sensible heat flux, soil heat flux, and drainage.

TERMINAL Section

The terminal section specifies the total simulation time FINTIM, the time interval for print-plotting (OUTDEL), the minimum tolerable time-step in the variable-step integration (DELMIN), and the list of variables to be printed or print-plotted against time.

D. Simulation Test of the Effect of Albedo Changes on Non-Isothermal Evaporation

The critical importance of soil surface conditions, particularly during early stages of evaporation, has long been recognized. Simulation now makes it possible to study the quantitative effect of various factors, including controllable changes in soil surface properties. Recent work (Jackson *et al.* 1974) has drawn attention to the changes of albedo which occur naturally during evaporation. Since albedo is one property which can readily be modified artificially, it is of interest to establish how and to what extent it might influence evaporation. An increase in albedo can, in principle, decrease evaporation by reducing the energy load on the surface and hence the temperature of the evaporation zone. On the other hand, a decrease of albedo, while causing greater warming of the soil surface, might also help to drive water vapour down into the profile and to arrest the evaporation process earlier by the more rapid desiccation of the surface. To answer the question adequately, therefore, a comprehensive modeling study, capable of handling soil water and energy dynamics simultaneously, is warranted. The model presented above has this capability.

To illustrate the operation and potentialities of the model, a 10-day model experiment was conducted in which three treatments affecting soil surface albedo were simulated: (1) leaving the soil surface in its "natural" state, with the albedo varying from a minimum of 0.1 to a maximum of 0.35 as the volumetric wetness of the surface soil decreased from 35% to 10% or less; (2) darkening the soil surface, as if by the application of charcoal powder, to produce a low albedo value of 0.05; (3) whitening the soil surface, as if by the application of chalk powder, to effect a high albedo of 0.50. The soil used in the simulation was assigned properties similar to those of Gilat fine sandy loam, and the climatic inputs were arbitrarily chosen to represent a typical late spring or early summer period in a semi-arid region (see Figure 2.2).

Figure 2.2a shows the course of evaporation during the ten diurnal cycles simulated. It is seen that the pattern of evaporation repeated itself during the first two days, but thereafter began to decrease, particularly for the low albedo conditions. The differences among the three treatments subject to the same meteorological inputs, seem to refute the simplistic assumption (used in our isothermal model) that the soil is merely a passive recipient of some external forcing function called "evaporativity" or "potential evaporation." In fact, the soil participates in determining its evaporation rate even before its surface becomes "air dry," inasmuch as its albedo and thermal conductivity affect the partitioning of energy and the fraction of the energy balance available for evaporation. Potential evaporation is thus seen to be a function of meteorological inputs and of soil properties jointly. Each soil condition seems to exhibit its own "potential evaporation" rate.

Figure 2.2b shows that the low albedo condition resulted in about 30% more evaporation in comparison with the high albedo condition during the first three days. The difference between the two albedo conditions later appeared to decrease so that after ten days, the cumulative evaporation value of the low-albedo soil was only about 8% greater than that of the high-albedo soil. It thus appears that increasing albedo results in reduction of evaporation only in the short run. It is noteworthy that the evaporation curve for the variable albedo condition (designed to simulate the pattern for a natural soil) at first resembled that for the low albedo condition, but as the soil surface dried and albedo increased, the pattern of evaporation tended to approach that of the high albedo condition. In any case, the curves for all three conditions tended to converge after the first few days.

Figure 2.3 shows the soil profile moisture content distribution at the end of 1, 2, 4, 6, and 10 days under the low albedo treatment (solid line curves) and the high albedo treatment (dashed line curves). Again, we see that the stronger and more rapid desiccation of the soil surface under the low albedo treatment was most pronounced after the first two to four days, but that the differences between albedo treatments seemed to disappear toward the end of the 10-day simulation period.

Increase of albedo is thus shown to be effective only insofar as it reduces the heat load on the surface and thus the potential evaporation, which is the primary determinant of evaporation rate during the initial stages of the process. In time, potential evaporation and soil surface conditions become less important as the transmission and supply of water from the deeper layers toward the surface become the dominant factor.

Figure 2.4 shows the day-to-day course of soil temperature at the surface, in the seedbed (30 mm depth), and below the seedbed (120 mm depth) during the ten diurnal cycles for the low, high,

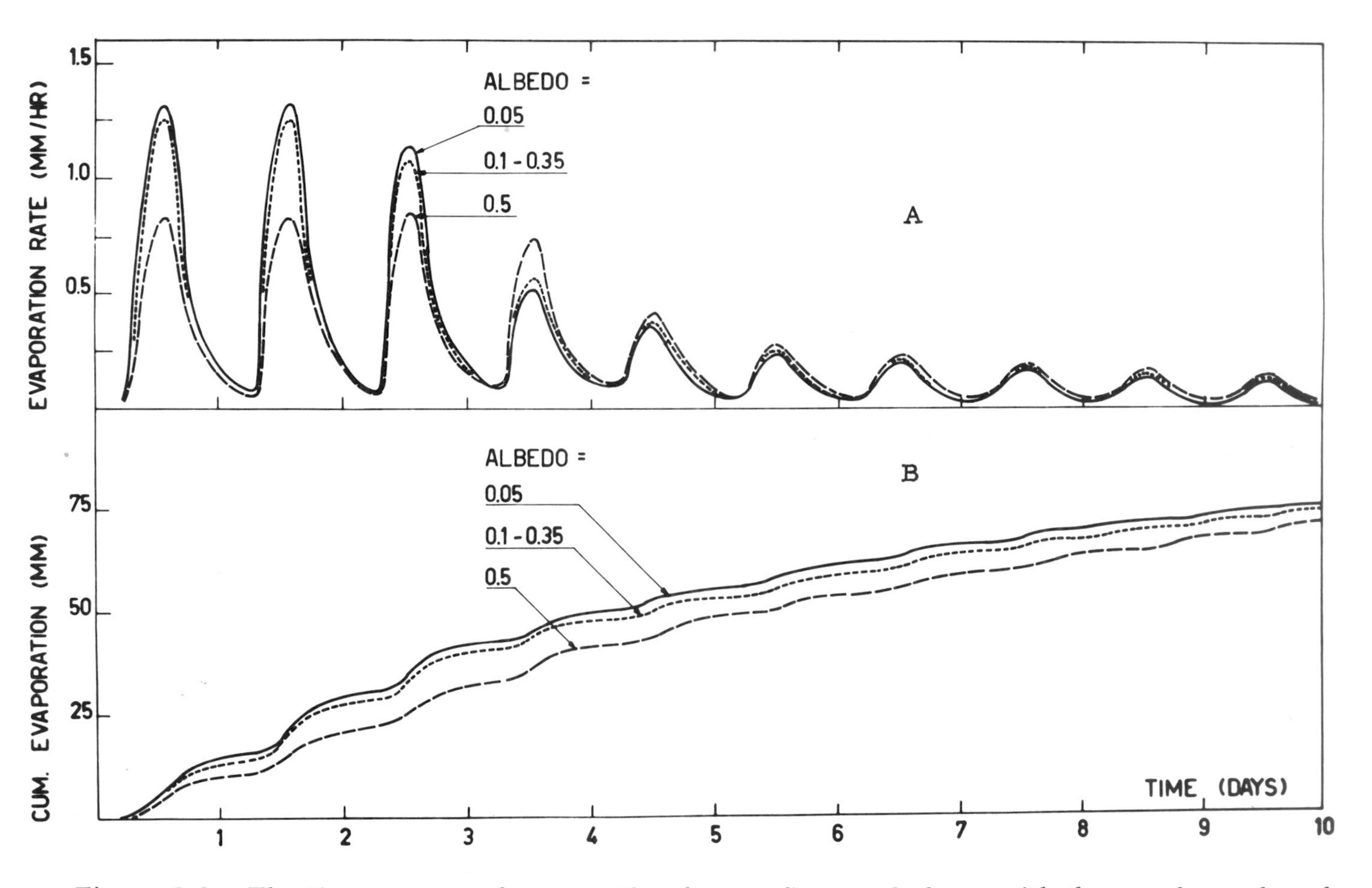

Figure 2.2. The time course of evaporation from a fine sandy loam with three values of surface reflectivity: low (0.05), high (0.5), and variable (0.1-0.35). The upper graph (A) indicates the daily fluctuation of evaporation rate, and the lower graph (B) shows cumulative evaporation during 10 days.

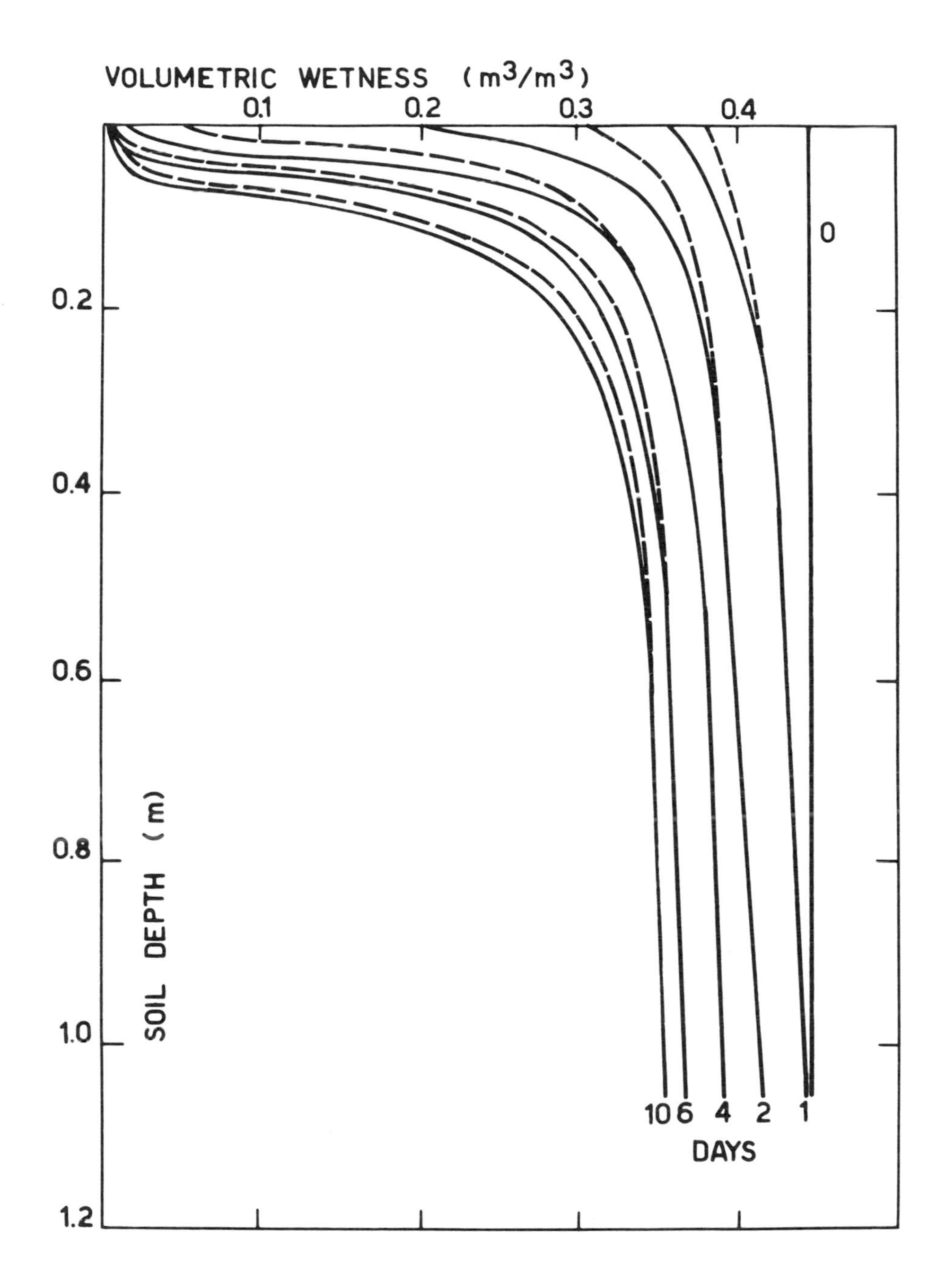

Figure 2.3. Soil profile moisture content distributions after 1, 2, 4, 6, and 10 days. Low albedo (0.05): solid-line curves. High albedo (0.5): dashed-line curves.

and variable albedo conditions. In all cases, the mid-day temperature maximum remained below 25°C during the first two or three days, during which evaporation took place at or about the potential rate for each condition. After the third day, however, as the soil surface tended to desiccate and the evaporation rate began to fall, thus consuming less energy as latent heat, the maximal surface temperature began to rise progressively to values approaching 40°C for the low albedo treatment and to about 30°C for the high albedo treatment, with the variable albedo again being between the two. The overall warming trend was evident in the progressive increase of the nightly temperature minimum as well as of the daily maximum. Comparison of the daily temperature waves for the various depths shows both a phase lag and a decrease of amplitude with increasing depth, phenomena which are too well known to require explanation.

Figure 2.5, finally, shows the temperature profiles at mid-day and at midnight of the 2nd, 6th, and 10th days for the three albedo conditions. Both the warming trend and the gradual increase of amplitude with time are evident. At the surface, the daily minima increased from about 11°C during the second day to about 18°, 19°, and 20° for the high, intermediate and low albedo conditions, respectively, whilst the corresponding daily maxima increased from about 18°, 20°, and 22° for the same treatments during the second day to about 30°, 34°, and 38° during the tenth day. Within the profile, the daily temperature wave and the warming trend penetrated to increasing depth, eventually reaching down below 80 cm, particularly under the low and variable albedo surfaces.

All results considered, we can conjecture that the principal potential benefit to be derived from modification of surface albedo are likely to be in affecting seedbed conditions during a germination period lasting several days. Where the weather is warm enough so that low temperatures are not likely to inhibit germination but soil moisture may be limiting, whitening the surface may conserve enough water in the seedbed to enhance germination appreciably. On the other hand, under cool weather conditions, where not moisture but low temperature may be the limiting factor, darkening the soil surface may be beneficial in hastening germination despite the increase of evaporation. However, this hypothesis, like all others resulting from modeling studies, remains to be tested in the field before it can be generally accepted.

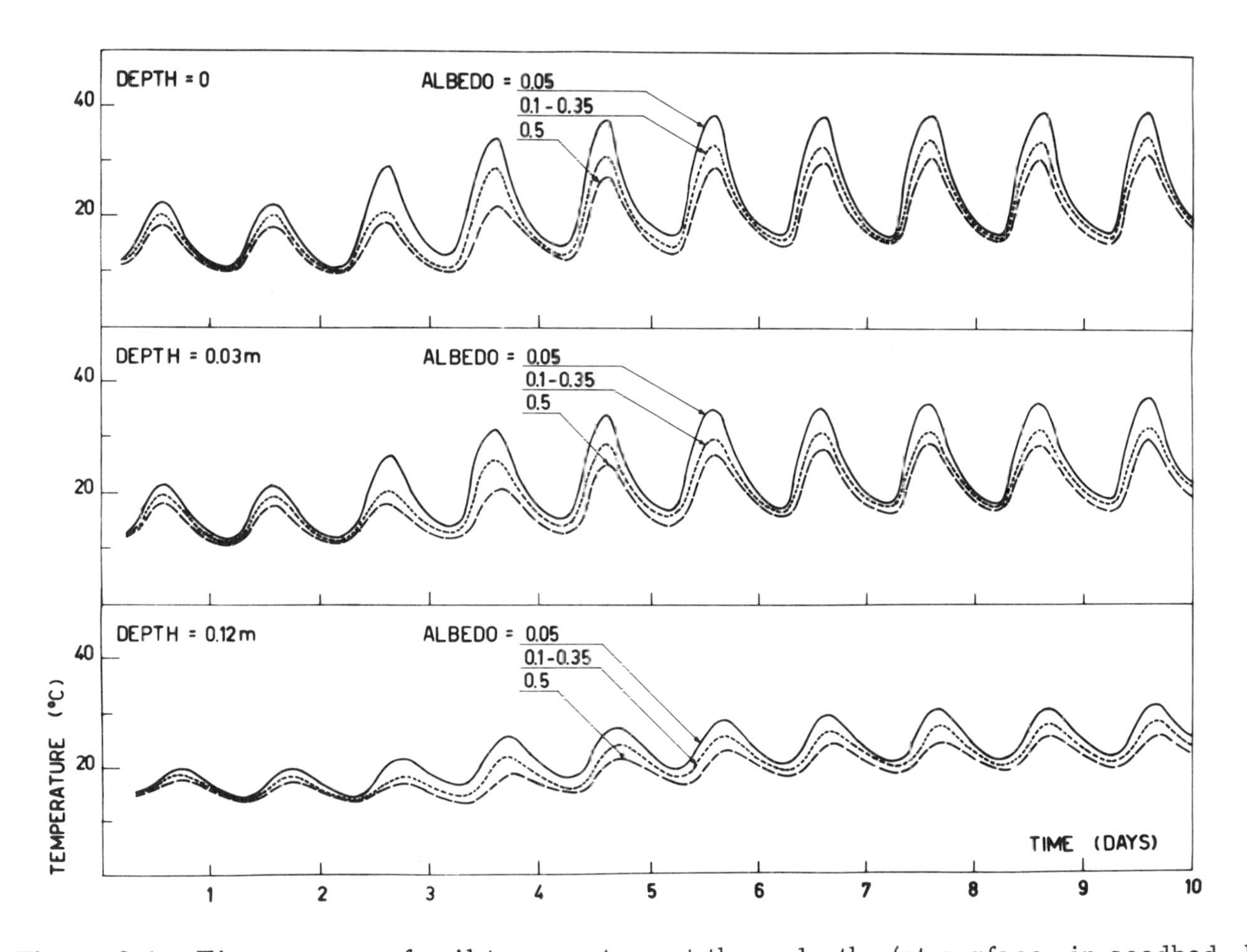

Figure 2.4. Time course of soil temperature at three depths (at surface, in seedbed, below seedbed) for three values of surface reflectivity (low, high, variable).

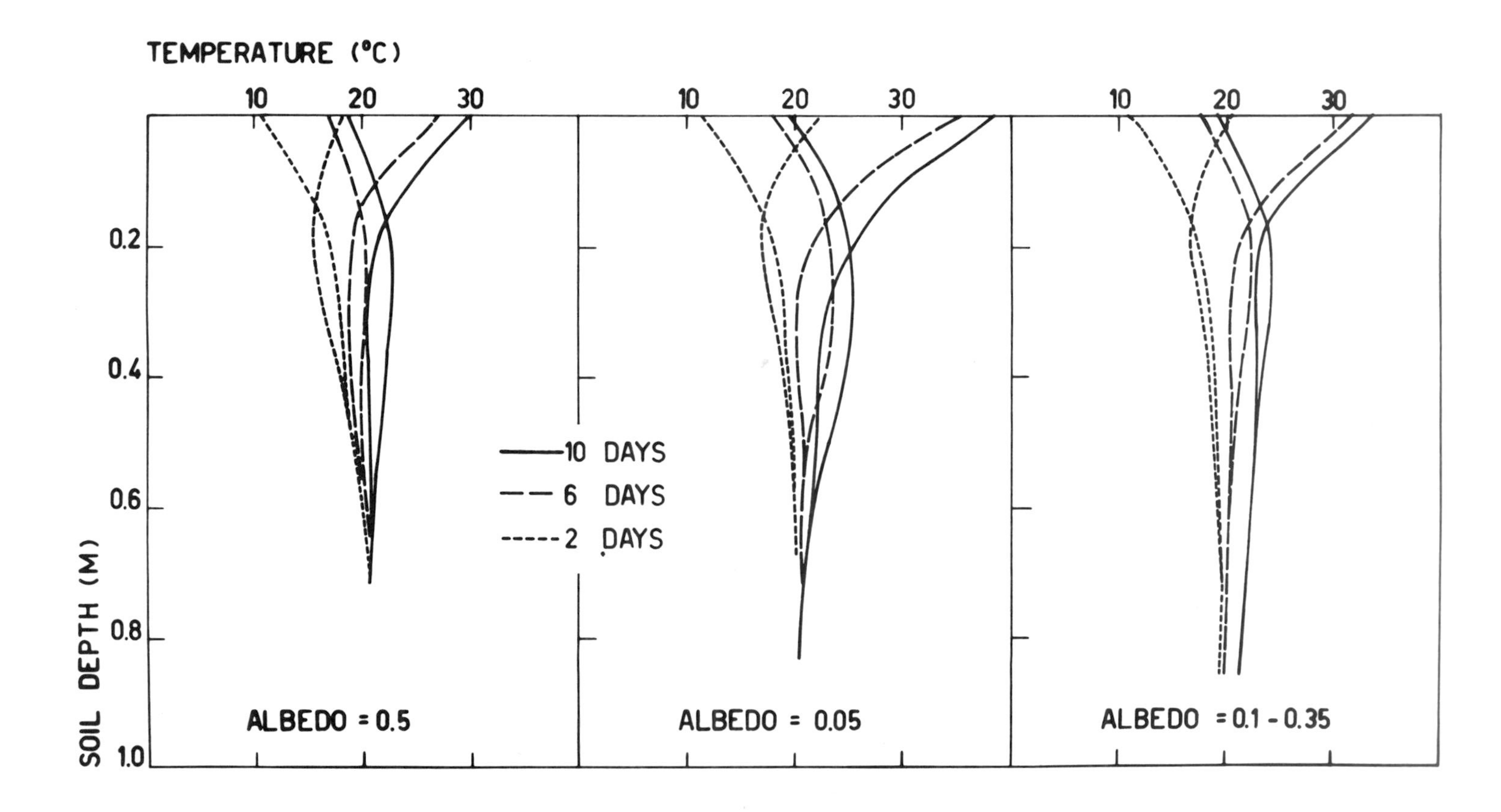

Figure 2.5. Soil temperature profiles at mid-day and at midnight of 2nd, 6th, and 10th day of simulated evaporation under low, high, and variable albedo conditions.

III. WATER DYNAMICS AND STORAGE IN FALLOW SOILS AS AFFECTED BY SOIL TEXTURE AND PROFILE LAYERING

A. Description of the Problem

Having devoted the first two case studies of this monograph to modeling the evaporation process, we now wish to widen our scope to the overall problem of soil water storage and conservation. In this larger context, evaporation is only one of several processes affecting the disposition of water in the soil profile, the other major processes being infiltration, surface runoff, and deep percolation.

Storage of water in the soil is a primary concern in the management of agriculture in arid and semiarid regions. Through repeated cycles of infiltration and evaporation, much of the water applied to the soil surface by rain or irrigation may be lost by runoff (which also entails the hazard of erosion), by direct evaporation or transpiration of weeds, or by internal drainage beyond reach of crop roots. In dryland farming, such losses can deprive the crops to be grown of a major portion of the limited rainwater supply and might result in crop failure. In irrigated farming, such losses reduce the efficiency of irrigation and water use.

To increase the efficiency of soil and water management, it is necessary to evaluate the balance and storage of soil moisture and to predict quantitatively to what extent they might be amenable to various possible control measures. Continuous knowledge of the amount and distribution of water in the soil profile can aid in deciding whether and when to plant various crops or to provide supplementary irrigation, in anticipating crop yields and in assessing the rate of groundwater recharge.

Several models have been developed to account for soil-water flow processes (Rubin 1967; Nimah and Hanks 1973; van der Ploeg and Benecke 1974). Still desired, however, is a user-oriented computational method to provide continuous information on the soil moisture status for various climatic and soil conditions, and quantitative criteria for appraising the possible benefits which might be expected from proposed or alternative soil management methods. The model presented herein is based upon the work of Hillel *et al.* (1975a), Hillel and van Bavel (1976) and Hillel and Talpaz (1977).

B. Soil Texture in Relation to Water Storage

Soil water storage and subsequent availability to crops depends in the first place upon the textural composition of the soil profile. That soils can differ greatly in their capacity to absorb and retain water has been known qualitatively since ancient times. In fact, early recognition of the importance of soil texture as a primary determinant of soil-water relations led the forerunners of modern soil physics to invest considerable effort into the development of procedures for determining and characterizing particle-size and later pore-size distribution in different soil types (*e.g.*, Keen 1931; Baver 1940). In themselves, however, the various "indexes" which had been proposed for the characterization of soil texture or structure have yielded no direct functional correlation with hydraulic processes taking place in the soil. Eventually, the preoccupation of classical soil physics with soil texture *per se* came to be regarded among many modern soil physicists as rather futile.

More recently, however, the development of experimental and theoretical methods has made it possible to quantify functional relationships connected with soil texture and structure which had heretofore been perceived and described only in qualitative terms. In particular, the use of computer-based dynamic simulation techniques now enables us to carry out a systematic series of theoretical experiments designed to map out soil behaviour during the entire sequence of processes comprising the field water cycle under any pattern of climate or management.

As an effort in this direction, we have chosen to characterize the hydraulic behaviour of three hypothetical soils, representing a sand, a loam, and a clay. From the assumed typical water retentivity functions and the hydraulic conductivity values at saturation, we can compute the unsaturated hydraulic conductivity as a function of soil wetness and suction. We then simulate the processes of drainage, evaporation, and infiltration, both separately and in combination. From this series of simulations, we can attempt to assess how texture, insofar as it determines basic hydraulic properties, might affect the dynamics and storage of water in uniform profiles of sand-like, loam-like, and clay-like soils.

C. Characterization of Soil Hydraulic Properties

The assumed soil moisture characteristic curves (water retentivity) for the *sand, loam,* and *clay* are shown in Figure 3.1. From these fundamental relationships, we derived the corresponding hydraulic conductivity function for each soil. To do this on a physically consistent basis, we used the theory first developed by Childs and Collis-George (1950), later modified by

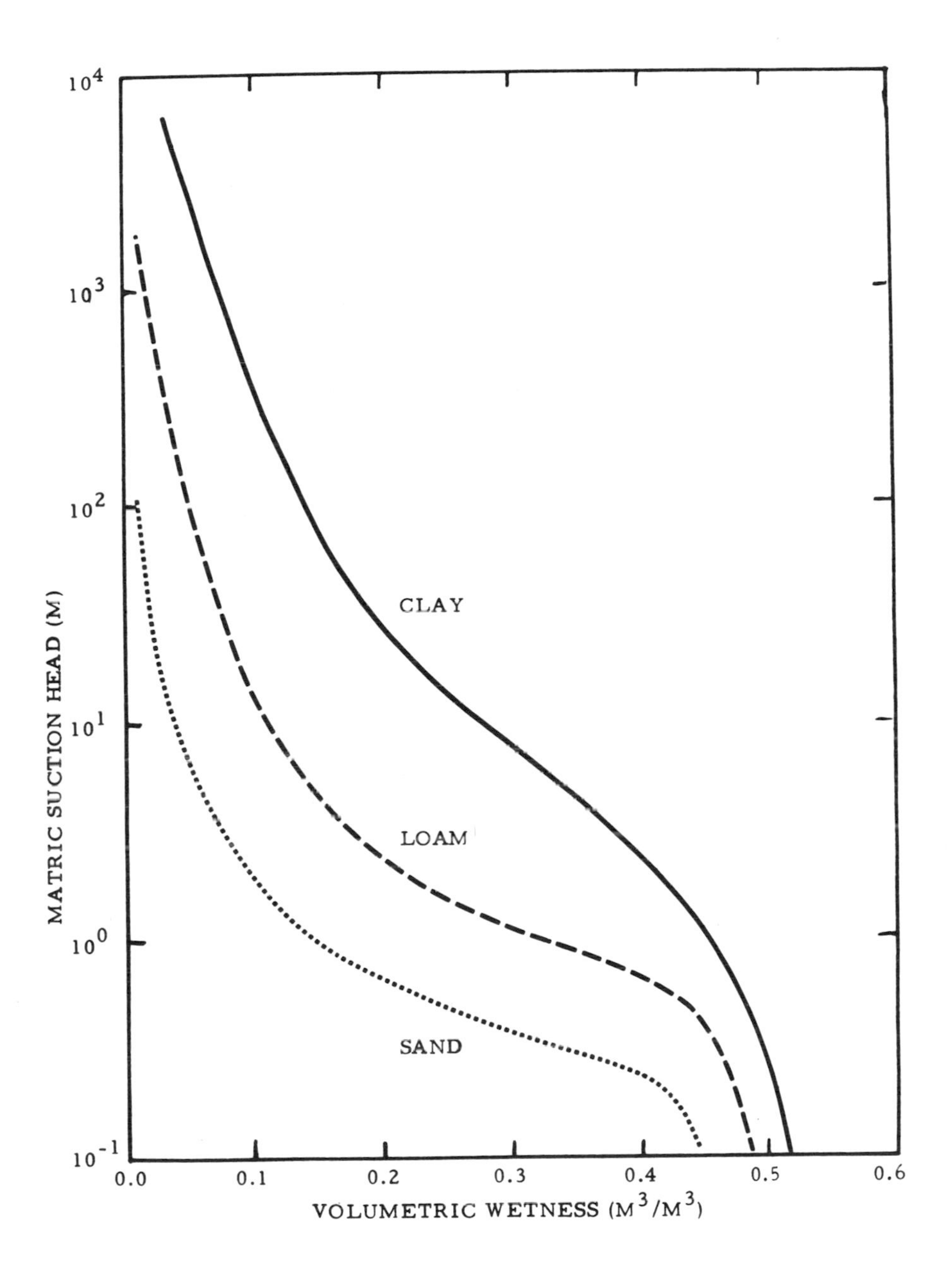

Figure 3.1. Soil moisture characteristic curves for the sand, loam, and clay used in the simulation of soil-water dynamics.

Marshall (1958) and Millington and Quirk (1959), and recently reformulated by Kunze *et al.* (1968) and Jackson (1972). In principle, this theory should make it possible to calculate the entire conductivity-wetness function, including the hydraulic conductivity at saturation. In practice, however, it was found to yield more realistic estimates when pegged to experimentally determined values of the hydraulic conductivity at saturation.

Following Jackson (1972), the conductivity function was obtained for each soil by dividing the soil moisture characteristic function into *n* equal wetness (θ) increments, determining the suction head (Ψ) at the midpoint of each increment, and calculating for each point a value of conductivity according to the equation:

$$K = K_S(\theta_i/\theta_S)^{\alpha} \frac{\sum_{j=1}^{m}[(2j + 1 - 2i)\Psi_j^{-2}]}{\sum_{j=1}^{m}[(2j - 1)\Psi_j^{-2}]} \qquad (3.1)$$

Herein, K_i is the hydraulic conductivity corresponding to any particular value of the soil's volumetric wetness θ_i; K_S is the hydraulic conductivity at saturation θ_S; Ψ is the suction head; α was taken by Jackson to be unity; i and j are summation indices; and m is the number of θ increments for which the calculation is to be made.

The respective θ_S and K_S values taken for each of our soils were as follows: 0.44 m^3/m^3 and 2.5 x 10^{-5} m/sec for the sand; 0.48 and 0.7 x 10^{-5} for the loam; and 0.52 and 0.2 x 10^{-5} for the clay.

The calculated hydraulic conductivities as functions of volumetric wetness are shown in Figure 3.2. From this relationship and the soil moisture characteristic (Figure 3.1) we plotted the conductivity vs. suction relation for each soil, as shown in Figure 3.3. The salient feature of these curves is the fact that they cross over, so that the relative conductivity of the various soils is different in the low and high suction ranges. This reversal will be seen to influence the patterns of water movement and storage in the various soil profiles during the infiltration-drainage-evaporation cycle.

D. Governing Equations for Processes Affecting Soil Water Storage

The various processes affecting water storage in soils of

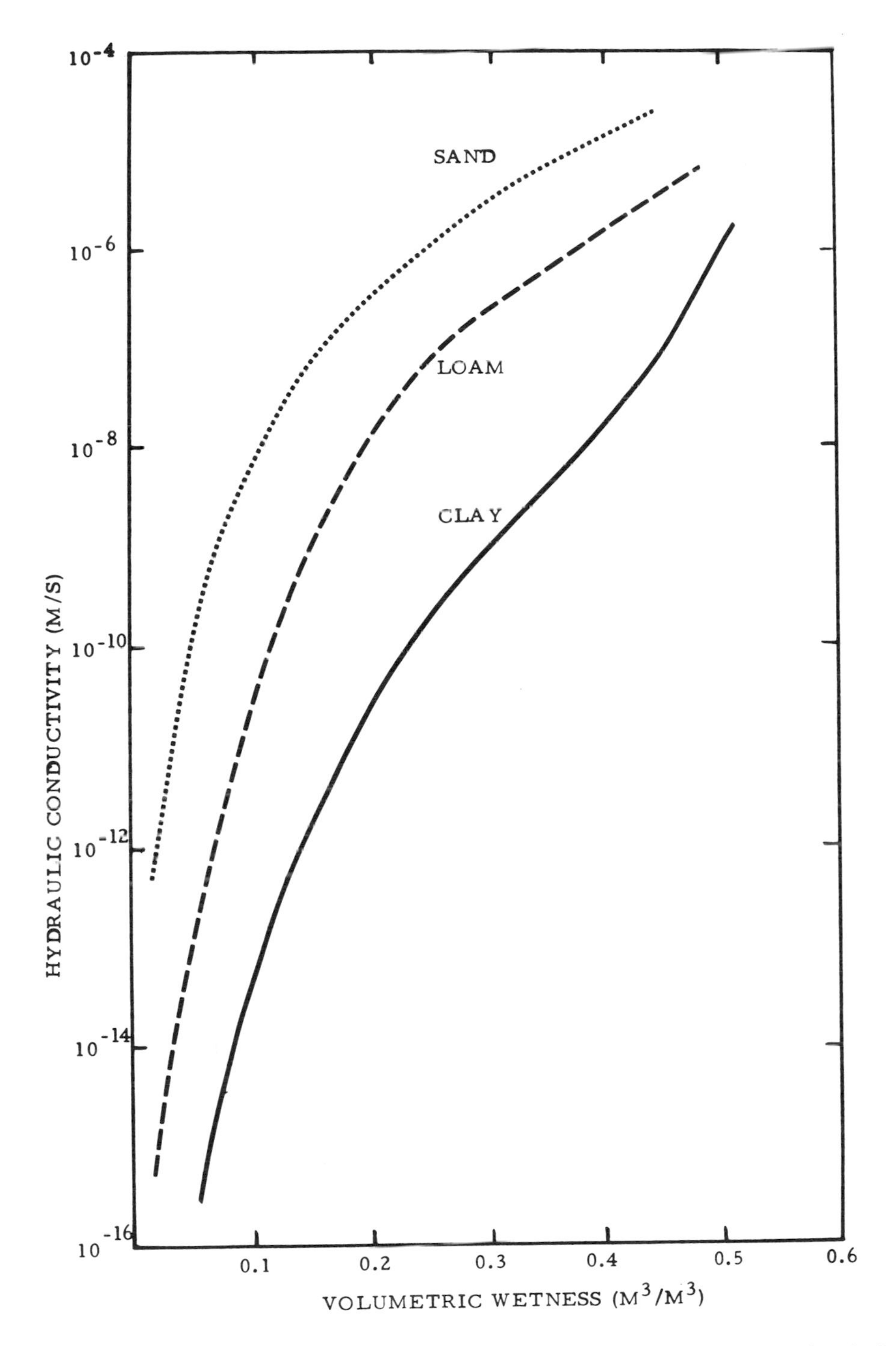

Figure 3.2. Soil hydraulic conductivity function derived from the soil moisture characteristic for each of the three simulated soil types.

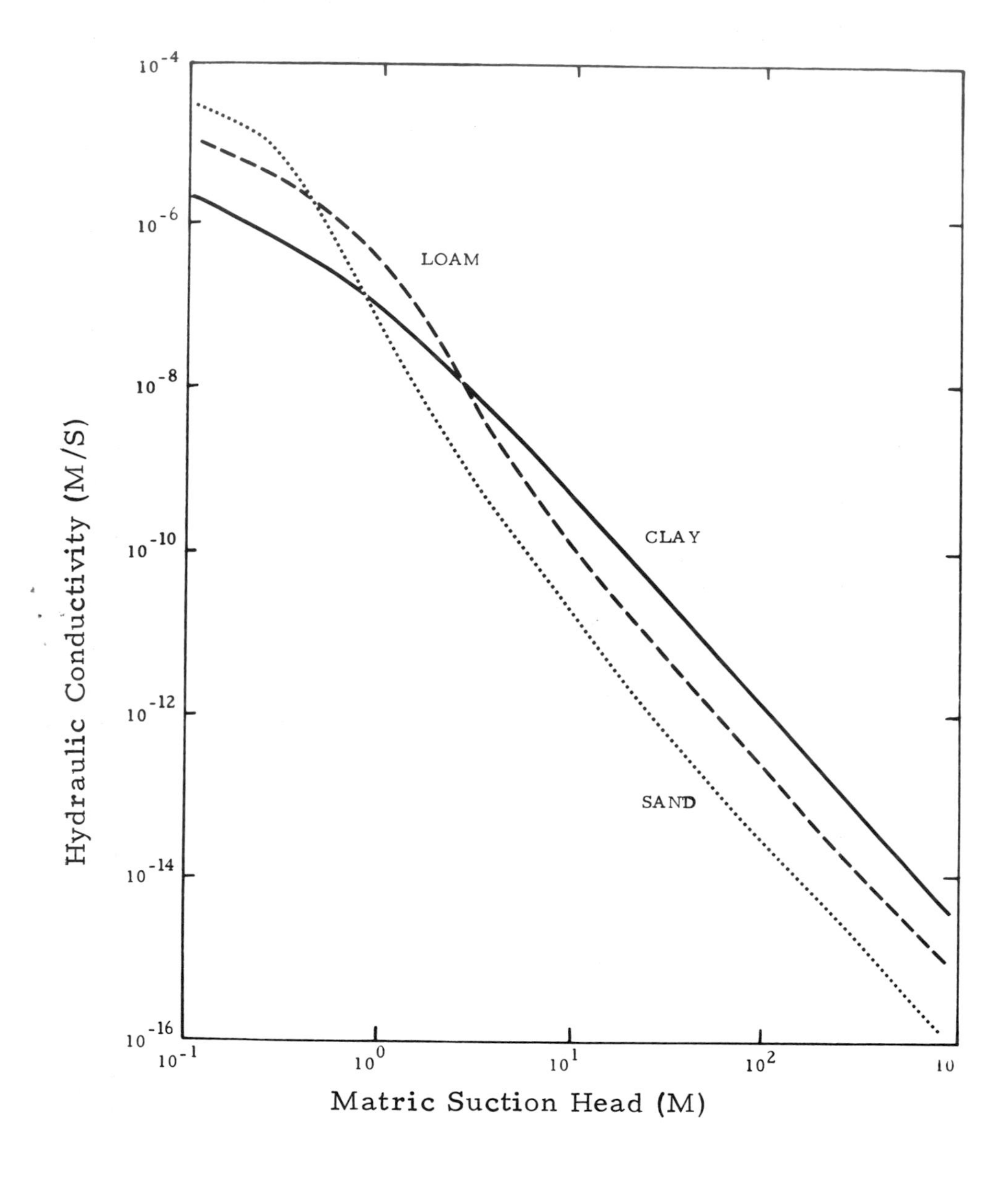

Figure 3.3. Hydraulic conductivity as a function of matric suction (hysteresis disregarded) for each of the three simulated soil types.

different textures will be considered first separately, and then in combination. These processes are:

1. For initially saturated profiles:
   a. Gravity drainage without evaporation.
   b. Evaporation under diurnally cyclic evaporativity without drainage.
   c. Simultaneous drainage and evaporation.

2. For initially unsaturated profiles:
   a. Infiltration under shallow ponding (infiltrability).
   b. Cycles of infiltration and evaporation simulating sequences of rainstorms and dry periods for different climatic or irrigation regimes.

To simulate gravity drainage without evaporation (case 1a) the profile was taken to be initially saturated. The flux through the upper boundary (the soil surface) was taken to be zero, while the downward flux through the bottom boundary (at a depth of 1.16 m) was specified as equal to the hydraulic conductivity of the lower layer (0.1 m thick). As the water content there decreased, the hydraulic conductivity and hence the transient rate of drainage diminished in time.

The process described consists of solving the vertical flow equations:

$$\frac{\partial \theta}{\partial t} = - \frac{\partial K(\theta)}{\partial z} - \frac{\partial}{\partial z}(K(\theta)\frac{\partial \Psi}{\partial z}) \tag{3.2}$$

subject to:

$$\begin{aligned} t &= 0, & z &\geq 0, & \theta &= \theta_s \\ t &> 0, & z &= 0, & q &= 0 \\ t &> 0, & z &= z_b, & q &= K(\theta_b) \end{aligned} \tag{3.3}$$

where t is time, z is depth, θ is volumetric wetness, θ_s is saturation, Ψ is matric suction, and K is hydraulic conductivity. Herein, z_b is bottom of the profile, θ_b is bottom-layer wetness, and q is the Darcian flux. The assumption of unit hydraulic gradient at the bottom of the profile (representing gravity drainage alone), though arbitrary, seems to be more realistic than the alternative conditions assumable for that zone, *e.g.*, constant flux, zero flux, or constant head. However, the model can be used with any other definable condition for the lower boundary.

To simulate evaporation without drainage (Case 1b), the profile was again initiated at saturation. The bottom boundary flux was set at zero and only upward flow was allowed. The evaporative flux through the soil surface was equal to the potential evaporativity (a forcing function) as long as the surface was moist. Thereafter, it equalled the upward flux from the profile into the dried top layer. The suction head of soil moisture at the top layer's midlevel, at a depth of 10 mm, was not allowed to rise above 1000 m, at which value the soil surface (depth zero) was assumed to have fallen to its "air-dry" value. This is equivalent to solving the horizontal flow equation (see Chapter I):

$$\frac{\partial \theta}{\partial t} = -\frac{\partial}{\partial x}\left(K(\theta)\frac{\partial \Psi}{\partial x}\right) \qquad (3.4)$$

subject to:

$$\begin{array}{llll} t = 0, & x \geq 0, & \theta = \theta_s & \\ t > 0, & x = x_b, & q = 0 & \\ t > 0, & x = 0, & \theta > \theta_d, & q = E_o \\ t > 0, & x = 0, & \theta \leq \theta_d, & q = K(\theta)\frac{\partial \Psi}{\partial x} \end{array} \qquad (3.5)$$

wherein x is the distance from the soil surface, and E_o evaporativity.

The pattern of evaporativity varied diurnally, following a sine function during daytime, as in the case study presented in Chapter I.

$$E_o = E_{max} \sin 2\pi/86400 \qquad (3.6)$$

where E_{max} is the maximal midday evaporation rate, taken to be 0.5 x 10^{-6}m/sec (1.8 mm/hour); and t is time in seconds from sunrise. Nighttime evaporativity was assumed to be steady at the rate of 1% of E_{max}. The total daily evaporativity was 13.7 mm.

To simulate simultaneous drainage and evaporation (Case 1c) Eq. (3.2) was solved subject to the same top conditions as in Eq. (3.5) and the same bottom conditions as in Eq. (3.3).

To calculate infiltrability, the moisture profile was initiated at a uniform suction head of 3.3 m, and the hydraulic head at the soil surface was set to zero. The time-dependent flux into the first soil layer, and the successive distributions of water in the soil profile, were calculated until the flux became steady. The solution to Eq. (3.2) was thus obtained subject to

$$\begin{array}{llll} t = 0, & z \geq 0, & \theta = \theta_i, & \Psi = \Psi_{1/3} \\ t > 0, & z = 0 & \theta = \theta_s & \Psi = 0 \end{array} \qquad (3.7)$$

where θ_s is the initial wetness value corresponding to a suction head of 3.3 m ($\Psi_{1/3}$).

Finally, to simulate repeated cycles of rain-infiltration and subsequent redistribution, evaporation, and through-drainage (Case 2b), the moisture profiles were initiated at a suction head of 30 m (*i.e.*, a pressure of -3 bars). The first rainstorm was begun at time zero, corresponding to midnight. Successive rainstorms were repeated at 2-day intervals. Each rainstorm lasted 6 hours: 2 hours of increasing intensity, 2 hours at a steady intensity of 0.5×10^{-5} m/sec (18 mm/hour), and 2 hours of diminishing intensity, for a total rainfall of 72 mm per rainstorm.

After the cessation of an infiltration period, and before commencement of the next, the soil surface was subjected to a diurnal pattern of evaporation, under cyclic evaporativity as per Eq. (3.6). Redistribution of soil moisture after each rainstorm occurred under the combined influence of gravity and pressure gradients, and depended upon the variable value of the unsaturated conductivity at each depth in the profile. Throughout the simulation, the bottom zone of the profile was assumed to be draining by gravity alone (*i.e.*, at unit hydraulic gradient) at a variable rate equal to the hydraulic conductivity of that bottom zone. Hydrologically, this lower boundary flux is the area's contribution to groundwater recharge.

E. A Conceptual Model of Surface Processes Affecting Soil Water Conservation

The principal logical components of our model, pertaining to surface processes affecting soil water storage, are presented in the form of a flow chart in Figure 3.4. The following processes are included:

(a) While Rain or Irrigation is in Progress

Evaporation is assumed to be nil. At the same time:

(1) If rainfall rate does not exceed the soil's infiltrability[1] and if no residual free water is present on the surface from previous rain, then the soil must be absorbing the rain as fast as it falls and the infiltration rate is taken to be equal to the rainfall rate.

(2) As infiltrability decreases or if rainfall rate is high

---

[1]*Infiltrability* is defined according to Hillel (1971) as the downward flux of water through the surface when the surface is maintained under a thin layer of water essentially at atmospheric pressure. As such, infiltrability is not constant but decreases as the hydraulic gradients decrease throughout the wetted portion of the profile. For an analysis of the formation of surface water excess during rain infiltration, see Swartzendruber and Hillel (1975).

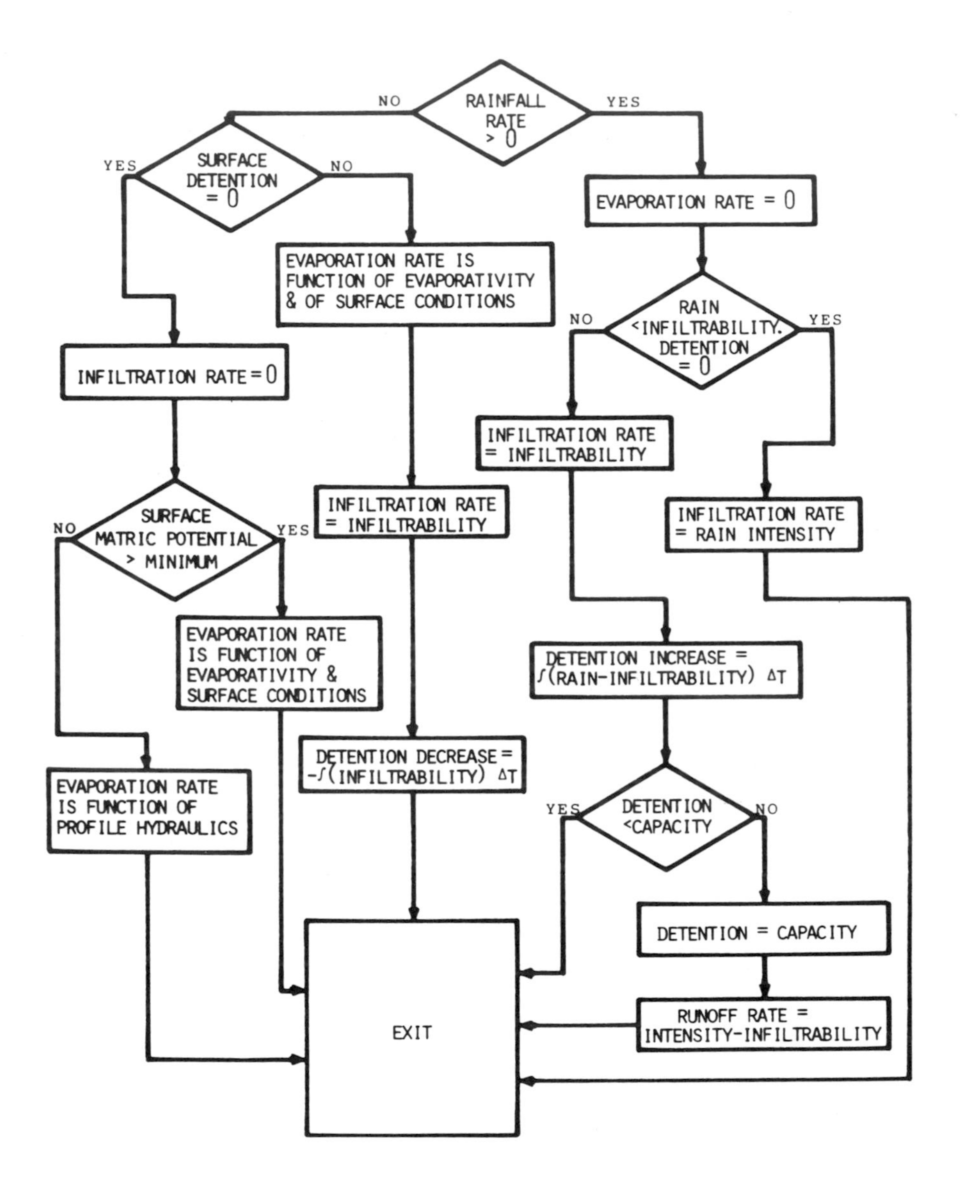

Figure 3.4. Flow chart indicating logical structure of the simulation model for processes occurring through the soil surface.

the latter may exceed infiltrability. Also, free water may be present over the soil surface from previous rain. In either case the infiltration rate is taken to be equal to the soil's infiltrability. If the soil surface is smooth and not entirely level, the excess rainfall rate over infiltrability can be expected to trickle downslope as surface runoff. On the other hand, if the surface is rough or pitted (or if it is covered by a mulch of hydrophobic clods) then the excess rain may accrue temporarily in puddles. Since the amount of water thus detained per unit area cannot exceed the surface storage capacity, continuation of the high rainfall rate will eventually form runoff.

(b) After Rain Has Ceased

(1) As long as free water is still detained over the soil surface, infiltration is assumed to continue at a rate equal to the soil's infiltrability. This infiltration gradually depletes the surface detention. At the same time, evaporation can take place, its rate depending on the potential evaporativity.

(2) If no free water is present over the soil surface, no infiltration can occur. Evaporation may take place, and cause a gradual drying of the soil surface zone. As long as that surface remains moist, evaporation proceeds at a maximal rate determined by external evaporativity and the effectiveness of the mulch as a diffusion barrier. However, as the soil surface dries down to some limiting value, the actual evaporation rate will no longer remain equal to the potential rate but must begin to fall below it. At this stage, the evaporation rate is no longer set by external and surface conditions, but by internal soil profile hydraulics which determine the flux of soil moisture delivered to the evaporation zone.

F. Description of the Computer Model

The program, also written in System 360 CSMP (IBM 1972), is presented in Figure 3.5. Many of the features in this program are similar to those of the preceding programs. Enough of the features are new, however, to warrant a complete (albeit partially repetitive) explanation.

INITIAL Section

Here the number of compartments (NL) is specified and their thicknesses tabulated (TABLE TCOM). The two top compartments were each made 0.02 m thick, and the underlying compartments were made to increase gradually in thickness for a total profile depth of 1.16 m. TABLE ITHETA again specifies the initial volumetric wetness, which was assumed to be at a uniform suction of 3 bars for all three soils.

An important provision of the program presented is the

possibility of dividing the soil profile into several layers of different textures. In the program given, the profile was divided into three textural layers: a toplayer consisting of five compartments having a total thickness of 0.16 m, an intermediate layer of five compartments with a total thickness of 0.28 m, and a subsoil layer which includes the remaining eight compartments (0.72 m). In principle, however, we can simulate any number or sequence of layers in a heterogeneous profile. On the other hand, to simulate a uniform rather than a layered profile, we can simply specify the same hydraulic properties (suction and conductivity functions) for all the layers.

The following parameters are assigned numerical values:

(1) SATCON: Hydraulic conductivity of the soil surface zone at saturation.

(2) DETCAP: Surface storage (detention) capacity in terms of average water depth stored in pockets or puddles at incipient runoff.

(3) PETMAX: Maximal potential evaporation rate (midday peak).

(4) MINPOT: Minimal matric potential head of surface compartment, representing air dryness.

Three pairs of tables are then given for the matric potential and hydraulic conductivity as functions of soil wetness, for each of the three soils, namely FUNCTION SUCTA and FUNCTION CONDTA for sand, SUCTB and CONDTB for loam, and SUCTC and CONDTC for clay.

The rainfall regime is given by FUNCTION RAINTB, which gives time (in seconds) versus rainfall rate (metres per second). The example given consists of two variable-intensity rainstorms, each lasting 6 hours and totalling 72 mm, the first commencing at the start of the simulation period and the second two days later.

DYNAMIC Section

The following calculations are made and updated at each time step during the simulation run:

(1) The volume of water in each compartment (VOLW) is calculated from the initial value (IVOLW) and the time integral of the net change.

(2) Evaporativity (PETO), by a sine function of time (diurnally repetitive).

(3) Volumetric wetness (THETA) of each compartment — from the ratio of the water volume to the compartment's volume.

(4) Matric potential (MPOT) and hydraulic conductivity (COND) for each compartment of specified texture — from the appropriate table given in the INITIAL section.

(5) Hydraulic potential (HPOT) is the sum of the matric potential head and gravitational head for each compartment.

(6) Hydraulic conductivity for flow between compartments

Figure 3.5. CSMP listing for the simulation of water storage in soil profiles of various textural composition—uniform or layered.

```
TITLE      SOIL MOISTURE REGIME IN UNIFORM AND LAYERED PROFILES
*       OF VARIOUS TEXTURES DURING CYCLES OF RAIN AND EVAPORATION

*          UNITS
*        KG = KILOGRAMS
*        M  = METERS
*        S  = SECONDS

*                         GLOSSARY OF SYMBOLS

* AVCOND = AVERAGE HYDRAULIC CONDUCTIVITY FOR FLOW BETWEEN COM-
*          PARTMENTS (M/S)
* BALANS = WATER BALANCE OF THE SOIL PROFILE (M)
* COND   = HYDRAULIC CONDUCTIVITY (M/S)
* CONDTA, CONDTB, CONDTC = HYDRAULIC CONDUCTIVITY TABLES FOR
*          SAND, LOAM, CLAY
* CUMDRN = CUMULATIVE DRAINAGE (M)
* CUMEVP = CUMULATIVE EVAPORATION (M)
* CUMINF = CUMULATIVE INFILTRATION (M)
* CUMPET = CUMULATIVE POTENTIAL EVAPORATION (M)
* CUMRNF = CUMULATIVE RUNOFF (M)
* DEPTH  = DEPTH OF MIDPOINT OF COMPARTMENT (M)
* DETAIN = DEPTH OF SURFACE-WATER EXCESS (M)
* DETCAP = DETENTION CAPACITY FOR SURFACE WATER EXCESS (M)
* DIST   = DISTANCE OF FLOW BETWEEN ADJACENT COMPARTMENTS (M)
* EVAP   = EVAPORATION RATE (M/S)
* FLUX   = FLOW RATE OF SOIL MOISTURE (M/S)
* HPOT   = HYDRAULIC POTENTIAL HEAD (M)
* I      = INDEX OF COMPARTMENTS (ORDINAL NUMBER)
* INCAP  = INFILTRATION CAPACITY (INFILTRABILITY, M/S)
* INFILT = INFILTRATION RATE (M/S)
* ITHETA = INITIAL VOLUMETRIC WETNESS (M3/M3)
* IVOLW  = INITIAL VOLUME OF WATER IN EACH COMPARTMENT (M)
* IWATER = INITIAL CONTENT OF WATER IN ENTIRE SOIL PROFILE (M)
* MINPOT = MINIMAL MATRIC POTENTIAL OF SURFACE COMPARTMENT (M)
* MPOT   = MATRIC POTENTIAL HEAD OF SOIL MOISTURE (M)
* NFLUX  = NET FLUX OF WATER INTO COMPARTMENT (M/S)
* NA     = NUMBER OF COMPARTMENTS COMPRISING UPPERMOST TEXTURAL
*          LAYER
* NAA    = ORDINAL NUMBER OF LOWEST COMPARTMENT IN SECOND LAYER
* NL     = NUMBER OF COMPARTMENTS COMPRISING THE ENTIRE PROFILE
* PETMAX = MAXIMAL (MID-DAY) POTENTIAL EVAPORATIVITY (M/S)
* PETO   = POTENTIAL EVAPORATIVITY (M/S)
* RAIN   = RAINFALL RATE (M/S)
* RAINTB = RAIN TABLE: TIME (S) VERSUS RAINFALL RATE (M/S)
* RUNOFF = RUNOFF RATE (M/S)
* SATCON = SATURATION HYDRAULIC CONDUCTIVITY OF SURFACE COMPART-
*          MENT (M/S)
* SUCTA, SUCTB, SUCTC = SUCTION TABLES FOR SAND, LOAM, CLAY
* TCOM   = THICKNESS OF COMPARTMENT (M)
* THETA  = VOLUMETRIC WETNESS OF SOIL (M3/M3)
* VOLW   = VOLUME OF WATER IN EACH COMPARTMENT (M)
```

```
STORAGE        TCOM(25),DEPTH(25),DIST(25),ITHETA(25),COND(25)
STORAGE        AVCOND(25),MPOT(25),HPOT(25),FLUX(25)
/      DIMENSION  THETA(25),VOLW(25),IVOLW(25),NFLUX(25)
/      EQUIVALENCE (VOLW1,VOLW(1)),(IVOLW1,IVOLW(1)),(NFLUX1,NFLUX(1))
FIXED          I,NL,NLL,NA,NAA,NAAA,N4A

INITIAL

NOSORT
PARAMETER NL=18,SATCON=.2E-5,DETCAP=.002,PETMAX=.5E-6,MINPOT=-1000.
TABLE TCOM(1-18)=2*.02,4*.04,4*.06,4*.08,4*.10
TABLE ITHETA(1-18)=5*.20,5*.02,8*.08
             NLL=NL+1
             NA=5
             NAA=NA+1
             NAAA=NA+5
             N4A=NAAA+1
             IWATER=0.
      DO 100 I=1,NL
             IWATER=IWATER+TCOM(I)*ITHETA(I)
             NFLUX(I)=0.
         100 IVOLW(I)=ITHETA(I)*TCOM(I)
             DEPTH(1)=.5*TCOM(1)
             DIST(1)=DEPTH(1)
      DO 110 I=2,NL
             DEPTH(I)=DEPTH(I-1)+.5*(TCOM(I-1)+TCOM(I))
         110 DIST(I)=.5*(TCOM(I-1)+TCOM(I))
FUNCTION SUCTA=.................
FUNCTION CONDTA=................
FUNCTION SUCTB=.................
FUNCTION CONDTB=................
FUNCTION SUCTC=.................
FUNCTION CONDTC=................
FUNCTION RAINTB=(0.,0.),(7200.,.5E-5),(14400.,.5E-5),(21600.,0.),...
      (172800.,0.),(180000.,.5E-5),(187200.,.5E-5),(194400.,0.)

DYNAMIC

NOSORT
             VOLW1=INTGRL(IVOLW1,NFLUX1,18)
             T=TIME-21600.
             PETO=PETMAX*AMAX1(.01*PETMAX,SIN(2.*3.1416*T/86400.))
      DO 200 I=1,NA
             THETA(I)=VOLW(I)/TCOM(I)
             MPOT(I)=-AFGEN(SUCTC,THETA(I))
             COND(I)=AFGEN(CONDTC,THETA(I))
         200 HPOT(I)=MPOT(I)-DEPTH(I)
      DO 205 I=NAA,NAAA
             THETA(I)=VOLW(I)/TCOM(I)
             MPOT(I)=-AFGEN(SUCTA,THETA(I))
             COND(I)=AFGEN(CONDTA,THETA(I))
         205 HPOT(I)=MPOT(I)-DEPTH(I)
      DO 207 I=N4A,NL
             THETA(I)=VOLW(I)/TCOM(I)
             MPOT(I)=-AFGEN(SUCTB,THETA(I))
             COND(I)=AFGEN(CONDTB,THETA(I))
         207 HPOT(I)=MPOT(I)-DEPTH(I)
```

```
      DO 210 I=2,NL
         210 AVCOND(I)=(COND(I-1)*TCOM(I-1)+COND(I)*TCOM(I))/   ...
                        (TCOM(I-1)+TCOM(I))
             FLUX(NLL)=COND(NL)
      DO 220 I=2,NL
         220 FLUX(I)=(HPOT(I-1)-HPOT(I))*AVCOND(I)/DIST(I)
             RAIN=AFGEN(RAINTB,TIME)
             INCAP=(0.-HPOT(1))*.5*(SATCON+COND(1))/DIST(1)
             DETAIN=INTGRL(0.,RAIN-INFILT)
             IF (RAIN.GT.0.) GO TO 350
             RUNOFF=0.
             IF (DETAIN.LE.0.) GO TO 330
             EVAP=PETO
             INFILT=INCAP
             FLUX(1)=INFILT
             GO TO 390
         330 INFILT=0.
             IF (MPOT(1).GT.MINPOT) EVAP=PETO
             IF (MPOT(1).LE.MINPOT) EVAP=AMIN1(PETO,-FLUX(2))
             FLUX(1)=-EVAP
             GO TO 390
         350 EVAP=0.
             INFILT=INCAP
             IF (RAIN.LT.INCAP.AND.DETAIN.LE.0.) INFILT=RAIN
             FLUX(1)=INFILT
             IF (DETAIN.LT.DETCAP) GO TO 390
             DETAIN=DETCAP
             RUNOFF=0.
             IF (RAIN.GT.INCAP) RUNOFF=RAIN-INCAP
         390 CONTINUE
      DO 320 I=1,NL
         320 NFLUX(I)=FLUX(I)-FLUX(I+1)
             CUMINF=INTGRL(0.,INFILT)
             CUMRNF=INTGRL(0.,RUNOFF)
             CUMEVP=INTGRL(0.,EVAP)
             CUMPET=INTGRL(0.,PETO)
             FLXNLL=FLUX(NLL)
             CUMDRN=INTGRL(0.,FLXNLL)
      DO 400 I=1,NL
         400 CUMWTR=CUMWTR+VOLW(I)
             BALANS=CUMWTR-IWATER-CUMINF+CUMEVP+CUMDRN
             CUMWTR=0.

TERMINAL

TIMER        FINTIM=864000., OUTDEL=10800., DELMIN=.1E-5
PRINT (Optional)
PRTPLT CUMINF, CUMRNF, CUMEVP, CUMPET, CUMDRN, CUMWTR, DETAIN, BALANS
METHOD RKS
END
STOP
```

is taken to be the average of the two compartments' conductivities (AVCOND), weighted according to their relative thickness.

(7) Flux between compartments (FLUX) obeys Darcy's law.

(8) Flux across the bottom boundary (FLUX(NLL)) is taken to be equal to the lowest compartment's conductivity.

(9) Rain intensity (RAIN) is obtained by interpolation from the rain table (RAINTB).

(10) Infiltrability (INCAP) is the flux into the first compartment from the saturated soil surface at which the matric potential is assigned a value of zero.

(11) Depth of water detained over the soil surface (DETAIN) is the time integral of the difference between rain intensity and infiltrability.

(12) If there is no rain but there is some detained surface water, evaporation rate (EVAP) is equal to the potential rate (PETO) and infiltration (INFILT) is equal to infiltrability (INCAP).

(13) If no water is detained over the surface, infiltration is zero, and evaporation is equal either to the evaporativity or to the upward flux of soil moisture toward the surface, depending on the matric potential of the surface soil (see statements following no. 330).

(14) If rain is occurring, infiltration rate is equal to either rainfall rate or infiltrability, according to the statements following no. 350.

(15) Cumulative values of infiltration (CUMINF), runoff (CUMRNF), evaporation (CUMEVP), potential evaporation (CUMPET), and drainage (CUMDRN) are then calculated by integration of the appropriate variable with respect to time.

(16) The water balance (BALANS) is summarized, as a check, by the following statement:

$$\mathrm{BALANS} = \mathrm{CUMWTR} - \mathrm{IWATER} - \mathrm{CUMINF} + \mathrm{CUMEVP} + \mathrm{CUMDRN} \quad (3.8)$$

wherein CUMWTR is cumulative water content of the entire profile and IWATER is initial total water content.

The remainder of the program deals with output. The time intervals and formats of the output are specified, whether the data are to be printed, plotted against depth, or print-plotted against time.

G. Simulation of Water Storage in Uniform Profiles

Results

Changes in soil moisture distribution of initially saturated uniform profiles of sand, loam, and clay during drainage, evaporation, and simultaneous drainage-cum-evaporation, are shown in Figures 3.6, 3.7, and 3.8. These profiles, representing successive moments in time (indicated in days) exhibit differences

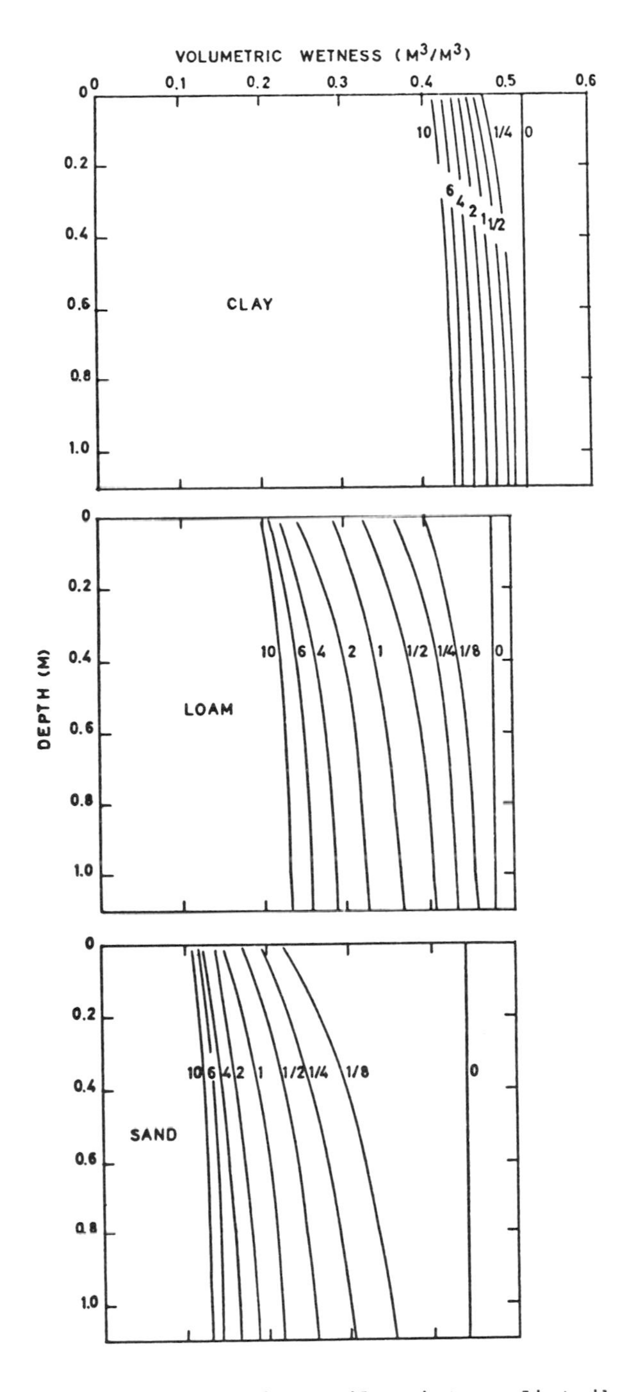

Figure 3.6. The changing soil moisture distribution during drainage from initially saturated uniform profiles of sand, loam, and clay.

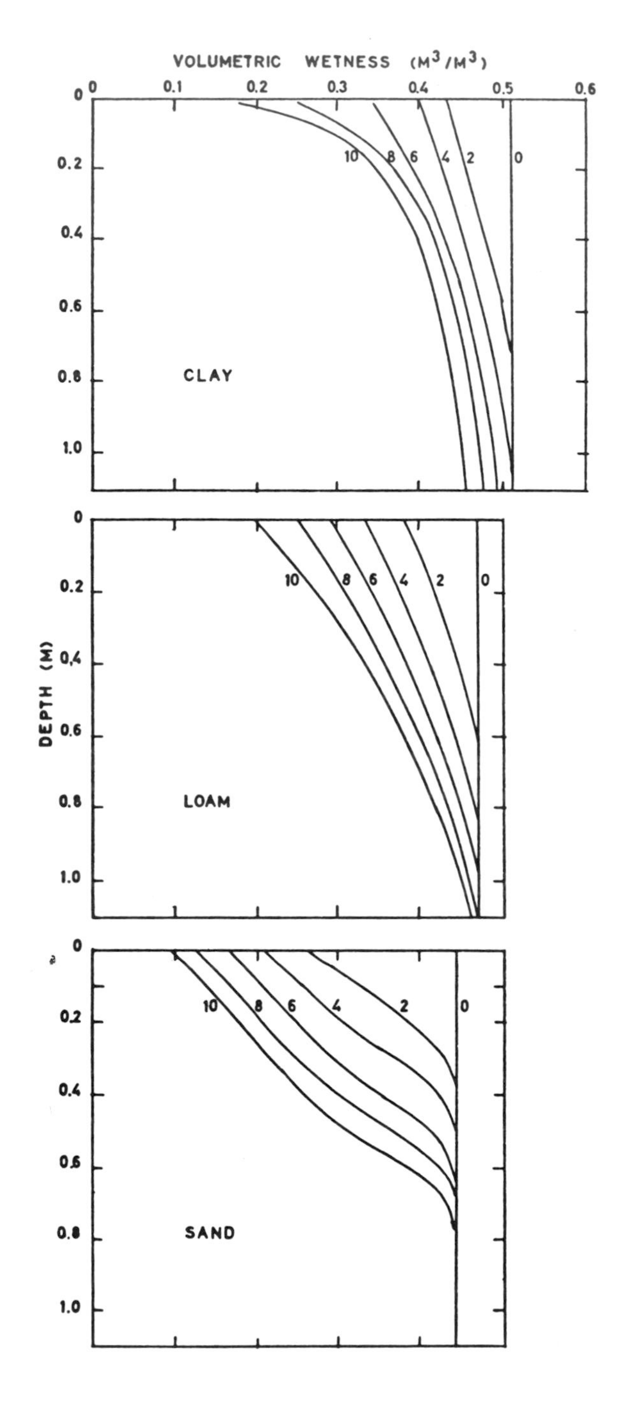

Figure 3.7. The changing soil moisture distribution during evaporation from initially saturated uniform profiles of sand, loam, and clay.

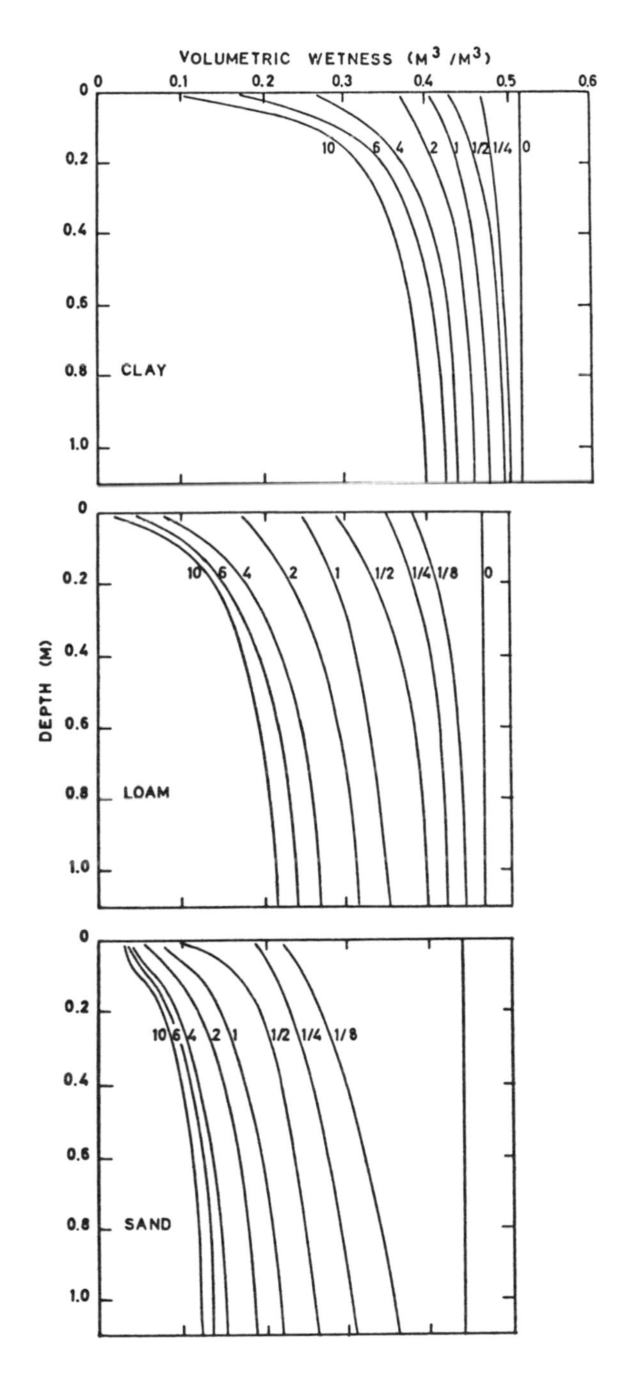

Figure 3.8. The changing soil moisture distribution during simultaneous drainage and evaporation from initially saturated uniform profiles of sand, loam, and clay.

in soil water dynamics among the various soils, caused by their differing hydraulic properties. A more detailed view of these differences is presented in Figures 3.9, 3.10, and 3.11.

Figure 3.9 shows the time-dependent rate of gravity drainage, with and without evaporation, out of the bottom of each of the 1.16 m deep simulated soil profiles. The sandy soil, although it contains less water at saturation, is seen to drain much more rapidly at first. Thus, it loses half again as much water as the loam and nearly five times as much water as the clay during the first two days. Thereafter, these differences decrease and are eventually reversed as further drainage from the sand slows down to a very low rate while drainage from the loam and the clay persists at appreciable rates for many more days. The evaporation process is seen to reduce drainage by a small but gradually increasing amount, and more so in the loam and clay than in the sand.

Figure 3.10 shows the cumulative evaporation loss from these profiles subject to an identical pattern of diurnally fluctuating evaporativity. It is seen that the sand sustained evaporation at the full potential rate only during the first day, whereas evaporation from the loam continued at the climatically determined potential rate for three days and evaporation from the clay persisted at this rate for as long as five days. This is apparently due to the higher water content of the clay as well as to its maintenance of higher unsaturated hydraulic conductivity (Figure 3.3). By the end of the 10-day period simulated, the simultaneously draining-cum-evaporating clay profile had evaporated more than twice as much as the sand, while the loam exhibited intermediate behaviour.

Figure 3.11 shows volumetric wetness at a depth of 0.41 m as a function of time in each of the three draining, and draining-cum-evaporating, profiles. This is pertinent to the "field capacity" problem. After two days of drainage from saturation without evaporation, the wetness values were about 45%, 29%, and 16% for the clay, loam, and sand profiles, respectively. Simultaneous evaporation reduced these values by about 2% in all three cases. These wetness values represent suction heads of 1.3 to 1.1 m for the clay and loam, and about 0.8 m for the sand. Thereafter, continuing drainage and evaporation tended to extract more water from this zone of the clay and loam profiles than from the sand, for a subsequent loss of an additional 8% or so on the tenth day from the clay and loam as against about 4% from the sand. The suction head values prevailing at this indicative depth after 10 days of internal drainage without evaporation were about 1.8 m in the clay and loam, and about 1.3 m in the sand. With evaporation, the corresponding values were about 4, 2.5, and 1.8 m for the clay, loam, and sand, respectively.

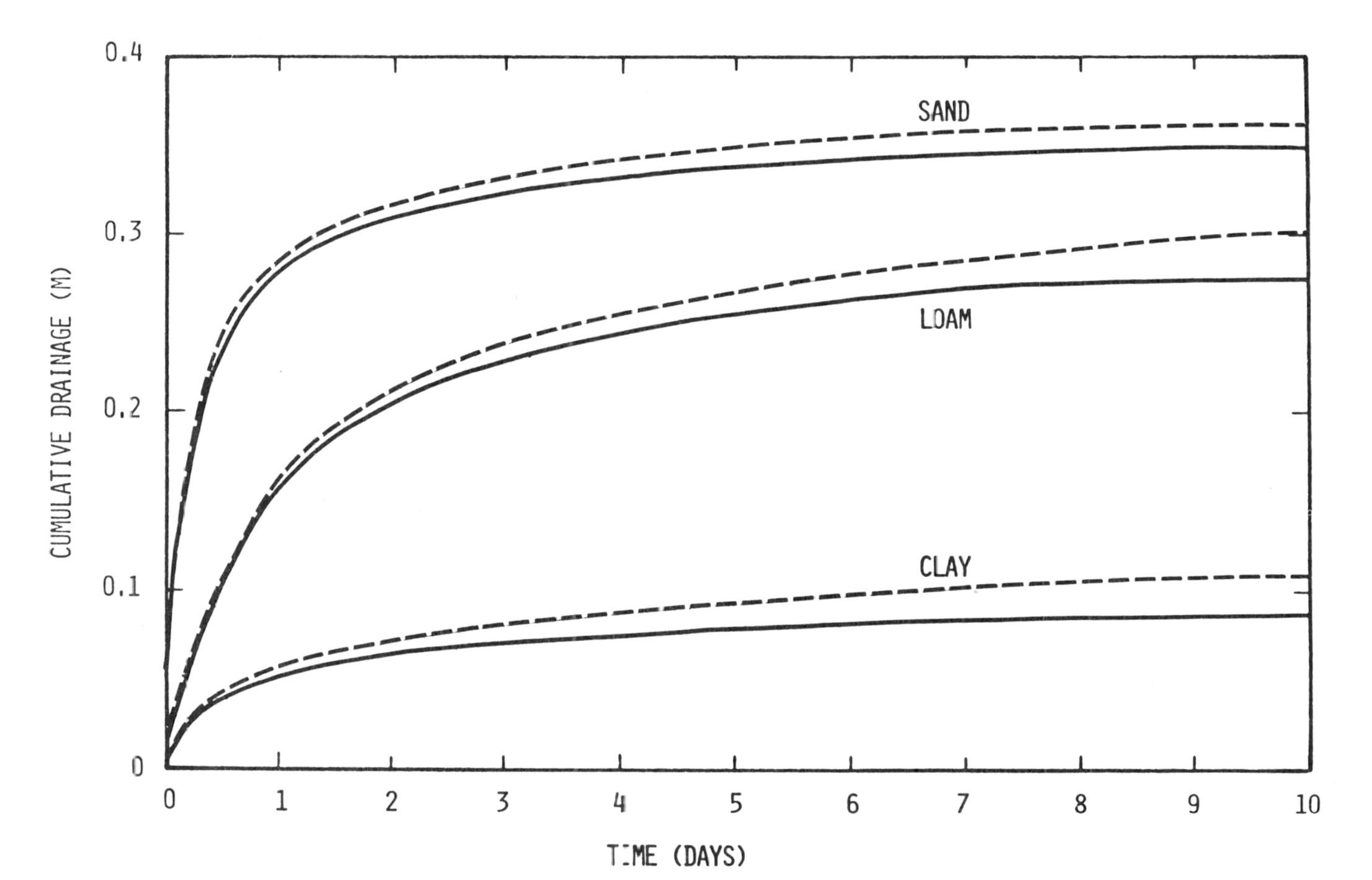

Figure 3.9. Cumulative drainage from initially saturated uniform profiles of sand, loam, and clay. Dashed lines: drain age without evaporation; solid lines: simultaneous drainage and evaporation.

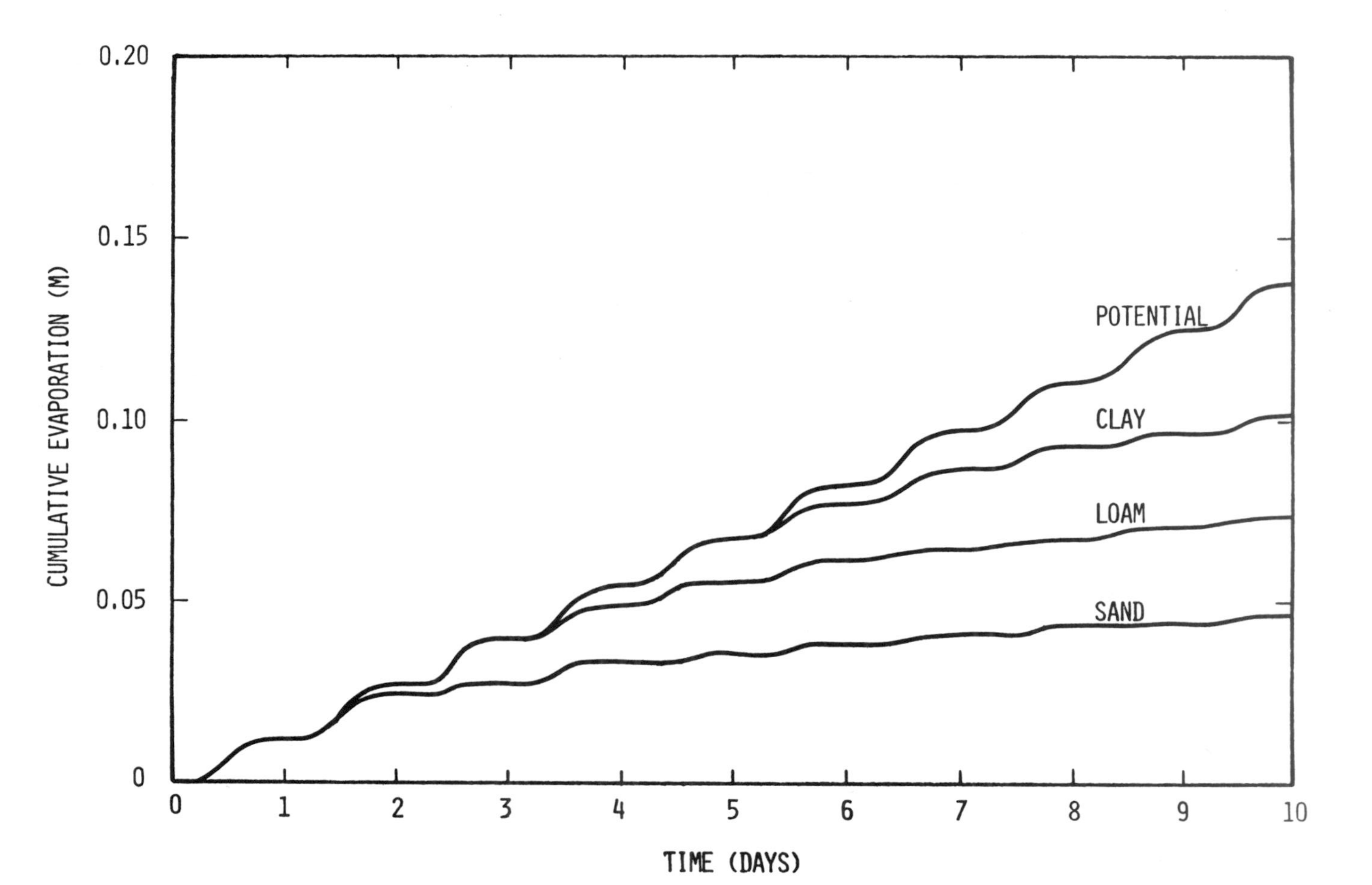

Figure 3.10. Cumulative evaporation during simultaneous drainage and evaporation from initially saturated uniform profiles of sand, loam, and clay.

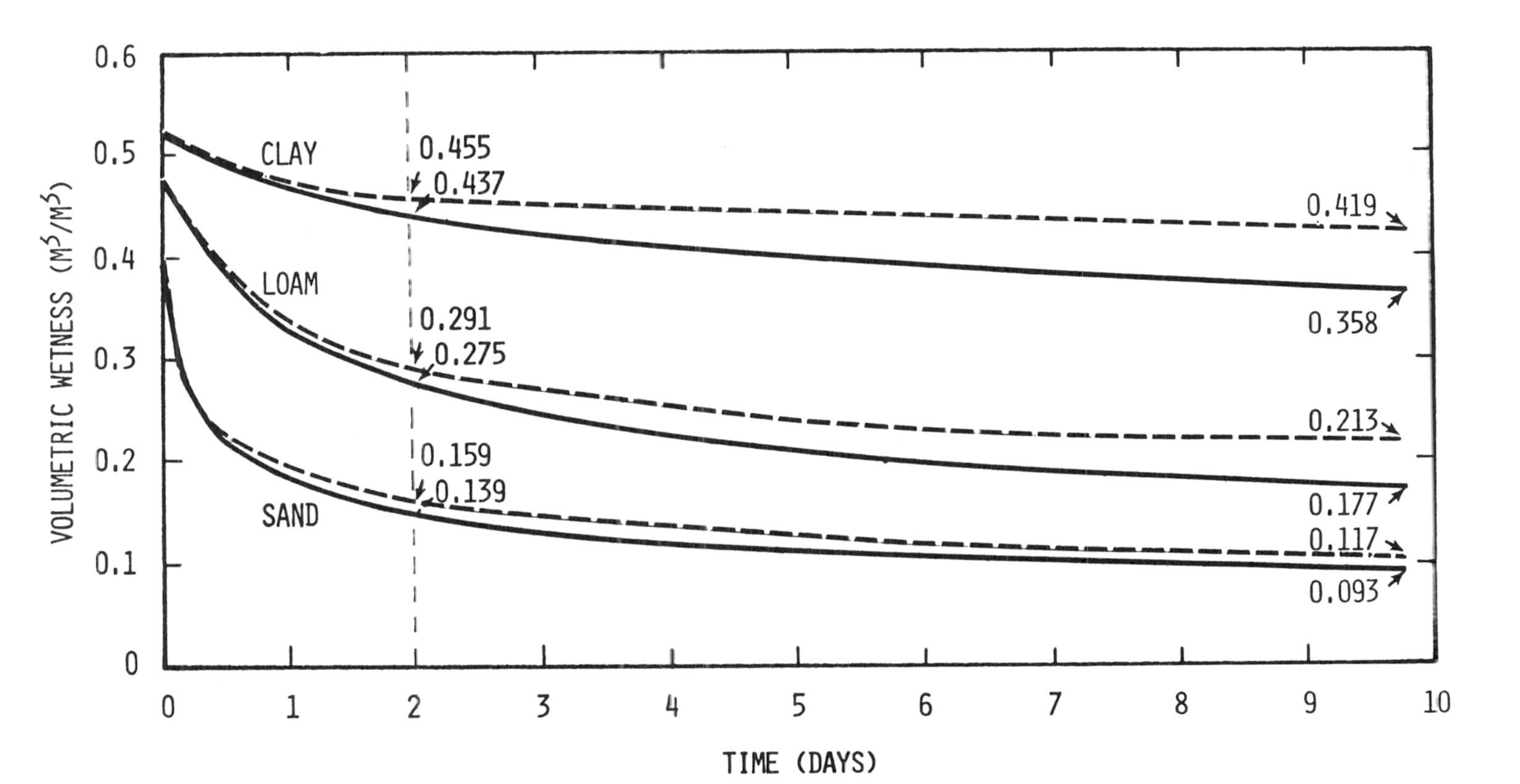

Figure 3.11. Volumetric wetness at depth of 41 cm as function of time in initially saturated uniform profiles of sand, loam, and clay. Dashed lines: drainage without evaporation; solid lines: simultaneous drainage and evaporation.

The course of infiltration into uniform profiles (Case 2a) of the three soil types is illustrated in Figure 3.12 where successive soil moisture profiles are plotted at various times as indicated by the numbers (signifying hours) inside the figure. The advance of the so-called "wetting front" is seen to be most rapid in the sand, even though its initial wetness was lowest, so that more water was required per unit volume to saturate this soil than the other two. Under simulated ponding, the sand profile was wetted throughout its entire 1.16 m depth within 4 hours, whereas the loam and the clay required 12 and 20 hours, respectively. Figure 3.12 also indicates that the "wetting front" was sharpest in the case of the sand, and least so in the clay, a phenomenon obviously related to the characteristic hydraulic functions in the pertinent suction range in each case.

The time course of cumulative infiltration under shallow ponding is illustrated in Figure 3.13. Here, the cumulative volume of water infiltrated into the soil profile per unit area of soil surface is plotted against the square root of time. The curves are seen to be straight at first, as the pressure gradients operating during the early phase of infiltration predominate over the gravitational force. However, with the pressure gradients decreasing as the profile wets up more and more deeply, gravity eventually dominates. As this occurs, the infiltration rate in a uniform profile approaches an asymptotic constant value which approximates the saturated hydraulic conductivity. On a linear scale of time, cumulative infiltration would become a straight line, whereas on the square-root of time scale this is indicated as a steepening curve. Figure 3.13 shows that at a $\sqrt{t}$ value of 150 (about 6 hours in actual time) the sand had absorbed about 650 mm, the loam about 250 mm, and the clay only about 70 mm.

Finally (Case 2b), we consider the results of simulating the processes of infiltration and of redistribution-cum-evaporation sequentially, under a programmed climate of successive rainstorms spaced by dry periods.

Successive soil moisture profiles during and following the first rainstorm (lasting 1/4 day) are shown for the three soils in Figure 3.14. In this case, the moisture profiles were initiated at a uniform suction head of 30 m. During the rainstorm, the upper zones of the clay and loam profiles approached saturation, as the maximal rainfall rate (0.5 x 10 m/sec) temporarily exceeded the saturated conductivity of the clay and nearly equalled that of the loam. The surface zone of the sand, however, remained unsaturated throughout the rain period, as the maximal rainfall rate was considerably lower than its saturated hydraulic conductivity.

Total penetration of water during this first rainstorm was deepest in the sand, but did not exceed 0.4 m in any case. Dur-

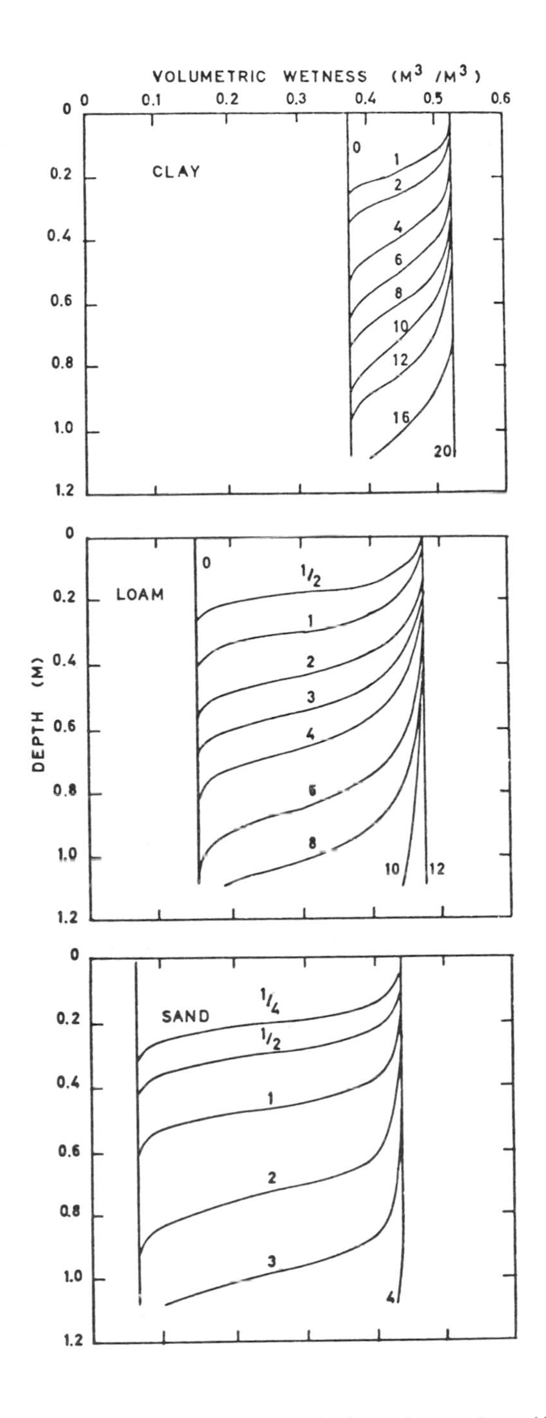

Figure 3.12. Successive distribution of soil moisture during infiltration into uniform profiles of sand, loam, and clay, all at an initial matric suction of 1/3 bar.

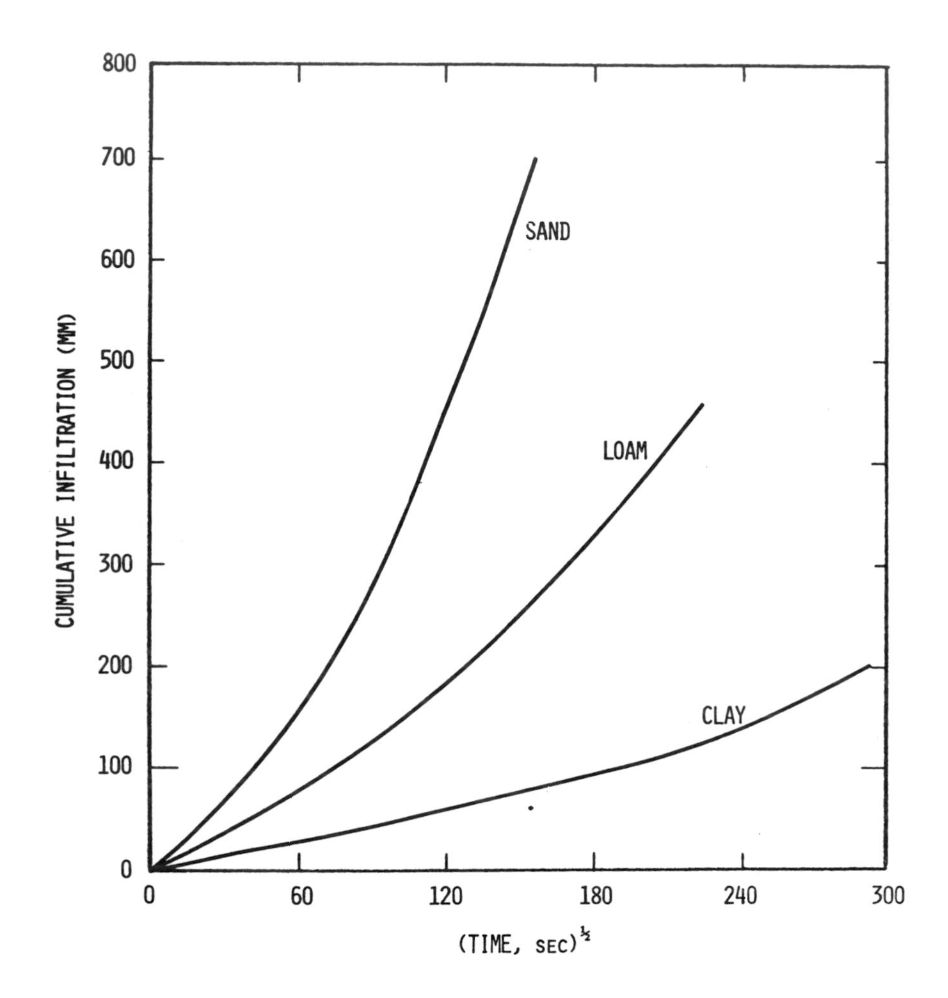

Figure 3.13. Cumulative infiltration into uniform profiles of sand, loam, and clay (all at an initial matric suction of 1/3 bar) as function of the square root of time.

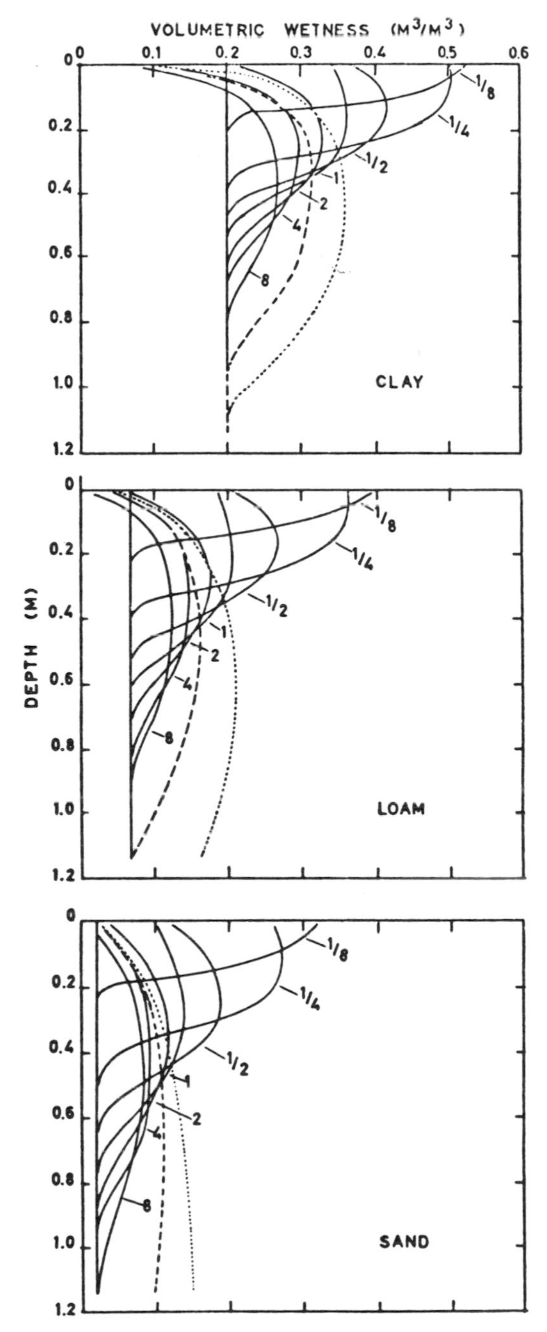

Figure 3.14. Successive distributions of soil moisture during infiltration and subsequent redistribution and evaporation in uniform profiles of sand, loam, and clay. The solid lines are for a single rainstorm of 72.2 mm lasting 6 hours, and the numbers indicate time (days) after onset of this rain. Dashed lines: soil moisture distribution at eighth day after three rainstorms totalling 216.6 mm.

ing the subsequent 8-day redistribution-cum-evaporation phase which followed this rainstorm, the penetration of water was shallowest in the clay (attaining < 0.8 m in depth), intermediate in the loam (about 0.85 m), and deepest in the sand, where penetration exceeded 1 m. In no case, however, did the single rain produce any appreciable through-drainage out the bottom of the 1.16 m profiles.

Figure 3.14 also suggests that, with most of the water remaining closer to the surface in the clay profile, the upward moisture-content gradients due to evaporation in the days following the single rain are steeper in the clay than in the loam and much steeper than in the sand profile. Observing the internal distributions of soil moisture after 8 days under the double-rain and the triple-rain regimes, we note that the bottom of the sand profile wetted up significantly under the double-rain, thus indicating appreciable through-drainage, whereas the bottoms of the clay and of the loam profiles were not wetted appreciably under this regime. Finally, we note that under the triple-rain regime both the sand and the loam profiles, but not yet the clay, were wetted to the bottom sufficiently to increase the conductivity there so as to indicate appreciable through-drainage.

We come at last to a summary of the total water balance of these 1.16 m profiles of sand, loam, and clay, under the various rainfall regimes simulated. This is shown in Table 3.1.

With a single rainstorm, there was neither runoff nor through drainage from any of the soil types. Hence, the only differences were in evaporation, the most occurring from the clay (50.3 mm) and the least from the sand (38.1 mm). Consequently, the sand profile stored 55% more water than the clay, as shown in the next-to-last column.

With two rainstorms, the sand profile already yielded some through-drainage, but not yet enough to be of any significance (1.2 mm). None of the soils yet produced any runoff. Hence, the major differences in the overall water balance were still due to evaporation. The clay still evaporated the most, and the sand the least, but the relative difference in the total amount of water added to storage was somewhat smaller, *i.e.*, the sand profile added 35% more than the clay.

With three rainstorms, the clay already produced appreciable runoff (3.6 mm) but the sand and the loam did not. On the other hand, the sand allowed considerable drainage (26.9 mm) and even the loam produced some (1.3 mm) but the clay none. The order of evaporation remained as before, but the differences decreased in magnitude. The overall balance of water storage, therefore, began to tip away from the sand (which added 101.1 mm) and toward the loam (which added 107.6 mm), with the clay profile storing almost as much (99.2 mm).

Table 3.1. Water balance of 1.16 m soil profiles of sand, loam, and clay under various rainfall regimes. 10-day simulation. Data in millimetres.*

Rainfall regimen	Soil texture	Total rainfall	Total in-filtration	Total runoff	Total evap-oration	Total drainage	Total water content			Storage efficiency
							Initial	Final	Added	
1 rain	Sand	72.2	72.2	0.0	38.1	0.0	23.2	59.3	34.1	47.2%
	Loam	72.2	72.2	0.0	43.1	0.0	81.2	110.3	29.1	40.3%
	Clay	72.2	72.2	0.0	50.3	0.0	232.0	253.9	21.9	30.3%
2 rains	Sand	144.4	144.4	0.0	66.3	1.2	23.2	100.0	76.9	53.3%
	Loam	144.4	144.4	0.0	78.7	0.0	81.2	146.9	65.7	45.5%
	Clay	144.4	144.4	0.0	87.3	0.0	232.0	289.1	57.1	39.5%
3 rains	Sand	216.6	216.6	0.0	88.6	26.9	23.2	124.3	101.1	46.7%
	Loam	216.6	216.6	0.0	107.7	1.3	81.2	188.8	107.6	49.7%
	Clay	216.6	213.0	3.6	113.8	0.0	232.0	331.2	99.2	15.8%
4 rains	Sand	288.8	288.8	0.0	108.4	66.9	23.2	136.7	113.5	39.3%
	Loam	288.8	288.8	8.2	126.8	25.2	81.2	209.8	128.6	44.5%
	Clay	288.8	270.7	18.1	130.9	1.6	232.0	370.2	138.2	47.9%

*Comment: The first rain occurred during the first night (before dawn of first day). The second, third, and fourth rains took place during the third, fifth, and seventh nights, respectively.

With four rainstorms, the balance of storage is reversed. Here, the differences in drainage become overwhelming, with the sand profile losing as much as 66.9 mm (almost 25% of the water infiltrated), while the clay allows only 1.6 mm to drain away. On the other hand, runoff losses from the clay are considerable (18.1 mm) yet not nearly as much as the sand loses to drainage. Differences in evaporation remain, but their relative magnitude decreases still further, so that the overall balance is now in favor of the clay, which indicates a storage increment of 138.2 mm as against 128.6 mm for the loam and only 113.5 mm for the sand.

It is noteworthy that the storage efficiency (*i.e.*, the amount of water added to the profile as percent of the total rainfall) generally varied in the range of 30% to 50%, with the highest value registered for the sand (under the double rain) and the lowest value for the clay (under the single rain). These numbers confirm the above conclusion that for small or moderate precipitation, the sand profile is the most efficient storage medium, whereas clay holds the advantage where rainfall is more abundant. It is remarkable that in no case was storage efficiency much greater than 58%. This is in accord with the experimental findings reported by Hide (1954).

Discussion

This model study of profile moisture dynamics in relation to soil texture and hydraulic properties was based on a rather arbitrary and hypothetical selection of soils and weather patterns. Hence we make no claim that our results are realistic in the sense that they can serve directly to describe any particular field situation. Our present model, furthermore, omits potentially important phenomena such as spacial heterogeneity, surface crusting or mulching, soil moisture hysteresis, energy relations (van Bavel and Hillel 1975), solutes, as well as the often dominant uptake of water by plant roots (Hillel *et al.* 1975c).

We chose to base our analysis on hypothetical rather than real soil properties because of the paucity of systematized data on the soil moisture characteristics of different textures, and the highly variable and often incomplete nature of the data that is available. However, our general familiarity with the available data leads us to believe that our curves are not untypcal of the textural classes designated, at least in their general shape.

Another problem we encountered was that matching incompatible suction and conductivity functions in a model of this sort can lead to computational difficulties. For this reason, we chose to derive each of the hydraulic conductivity functions from the corresponding soil's moisture characteristic. We are mindful however, that the theory used is based on the capillary hypothesis, which regards the conductive pores of the soil as if they were interconnected segments of cylindrical tubes. As such, this theory is applicable more to coarse-grained soils than to clay, where flow phenomena other than Poiseuille-type flow can be important.

One important factor which should eventually be included in the interest of realism is the hysteresis effect. For cyclic sequences of wetting and drying processes, the developing profiles with hysteresis taken into account are likely to be different from those presented. Specifically, hysteresis has been shown to retard redistribution (Rubin 1967) and cyclic evaporation (Hillel 1976). Moreover, the magnitude of the hysteresis effect is likely to be greater in sand than in clay. A quantitative simulation study of this topic, carried out by the author, does not, however, appear to contradict the basic relationships presented in this chapter.

In any case, we believe that a model study of the sort presented herein can aid in understanding the interacting functional relations governing soil moisture dynamics in varying circumstances. Even in its present, admittedly incomplete, form our model study carries certain significant plant-ecological and hydrological implications. Ecological investigations of arid regions (*e. g.*, Hillel and Tadmor 1962) have long shown that in the desert, sandy soils offer more favorable moisture conditions for plant growth than do finer-textured soils, thanks to the sandy soil's greater infiltrability and smaller runoff losses, deeper penetration of rain water due to lesser retention per unit depth, and lower evaporation losses. This is where rainfall is so scant that no appreciable percolation can occur beyond the reach of plant roots. In semihumid areas, however, the situation is often reversed in that sand constitutes the driest habitat owing to its excessively rapid drainage of the more abundant rainfall. Finally, in very humid areas the situation might again reverse itself as the frequent supply of water might be enough to maintain an adequate moisture level in the sand despite its high drainage rate while the storage of water in a clay soil could actually become excessive and impede the necessary root-zone aeration. These seeming vagaries can now be understood in quantitative terms with the aid of a simulation model capable of mapping out the entire gamut of interactions among the pertinent factors of soil and climate which enter into the determination of soil water storage and availability.

The hydrological implications of this sort of model study pertain to the possibility of *a priori* formulation of such hydrologically important factors as the initiation, rate, and quantity of runoff as well as of deep percolation (*i.e.*, ground water recharge) in terms of real physical mechanisms rather than *a posteriori* on the basis of statistical or empirical curve-fitting indexes. In fact, the need to introduce soil-physical processes into hydrology is one of the most challenging tasks of that vital science today, as it might help to bridge the gap between the traditionally separate fields of surface and ground water hydrology. An effort in this direction is made in our next chapter (Chap. IV).

H. Simulation of Water Storage in Texturally Layered Profiles

General

Numerous theoretical treatments of soil water dynamics have dealt with processes assumed to take place in ideally uniform soil profiles. Such studies, while valuable in themselves, can seldom if ever portray the behaviour of real soils, which, as a rule, vary in space and time, both texturally and structurally. In recent years, several investigators (*e.g.*, Hanks and Bowers 1962; Hillel and Gardner 1969; van Keulen and van Beek 1971) have attempted to define the behaviour of stratified soil with respect to separate flow processes such as infiltration. The advent of numerical simulation techniques now makes it possible to treat more complex and hence more nearly realistic cases by linking phenomena (including infiltration, redistribution, and evaporation) which have heretofore been considered separately. The treatment of such interacting phenomena as sequential or simultaneous processes in the context of a comprehensive model makes it possible to map out soil behaviour under various definable weather patterns.

This study is an outgrowth of a recent paper on soil moisture dynamics of layered profiles (Hillel and Talpaz 1977), providing an extension of the uniform profile model presented above.

Procedure

The same three soil types as in the preceding section (representing a "sand," a "loam," and a "clay") having the hydraulic properties shown in Figures 3.1, 3.2, and 3.3, were arranged in various vertical sequences to form 2-layer and 3-layer profiles. The layer depths were as follows: layer 1, 0-0.16 m; layer 2, 0.16-0.44 m; layer 3, 0.44-1.16 m. The hydraulic behaviour of these profiles was then explored systematically for specific boundary and initial conditions. The following processes were simulated over a period of 10 days:

(1) For initially saturated profiles:

(a) Gravity drainage without evaporation.

(b) Simultaneous drainage and evaporation under diurnally cyclic evaporativity.

(2) For initially unsaturated profiles:

(a) Infiltration under shallow ponding (infiltrability).

(b) Sequences of rain infiltration and periods of

drainage and evaporation.

The computations were based upon equations (3.2) through (3. 7). The same scheme (Figure 3.4) as in the simulations involving uniform profiles, as well as the same rainfall and evaporativity regimes, were used. A key assumption in this simulation is that in layered profiles, though the hydraulic conductivity and water content distribution may be discontinuous, the hydraulic potential of soil water is continuous throughout the profile even across layer interfaces.

Results

The soil moisture distributions of initially saturated two-layer profiles of all six possible combinations (*i.e.*, sand over loam, clay over sand, etc.) at the end of a 10-day period of gravity drainage are shown in Figure 3.15. For the period of time considered, it appears that a sublayer of sand can retard the drainage of top layers of clay and loam. This is apparently related to the rapid drainage of the underlying sand itself, which brings about a steep decrease of its hydraulic conductivity (as evident in Figure 3.3), thus arresting outflow from the overlying fine-textured layer. On the other hand, a fine-textured sublayer, which in itself drains more slowly at first but more persistently in the long run, can eventually cause greater outflow from a coarse-textured top layer. This pattern depends, of course, not only on the layering sequence but also on the layer depths and the period of time considered.

Cumulative drainage from the entire profile, including both top layer and sublayer, is shown for the various layering combinations in Figure 3.16. Since the sublayer of each profile in this two-layer simulation series was about six times thicker than the top layer, the overall cumulative drainage is dominated largely by the properties of the sublayer. Thus, the three profiles with clay sublayers are clustered at the bottom, and those with sand sublayers appear at the top of the family of curves shown. Within each group, moreover, the greatest cumulative drainage occurs where the top layer is sand, and the least where the top layer is clay.

The specific volumetric wetness near the bottom of the top layer (depth = 0.14 m), as it decreases during drainage, is shown in Figure 3.17. Once again, it is evident that the pattern of drainage from any top layer is influenced by the hydraulic properties of its sublayer, and since the latter vary in time, the relative pattern for different profile combinations also varies during the drainage process. Thus, for instance, soil moisture in a loam overlying clay decreases more slowly at first than in a loam overlying sand, but after four days or so the former catches up with the latter and its wetness continues to decrease at an appreciable rate through the tenth day. This reversal is due,

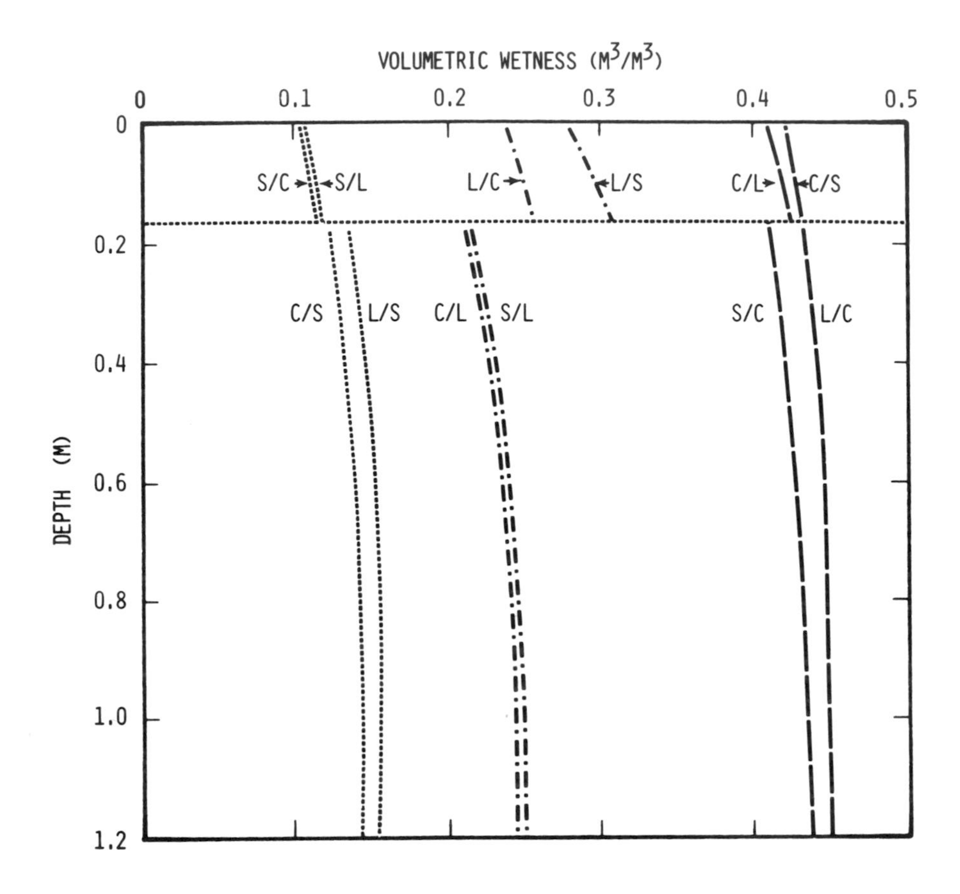

Figure 3.15. Moisture distributions in 2-layer profiles at the end of a 10-day drainage period.

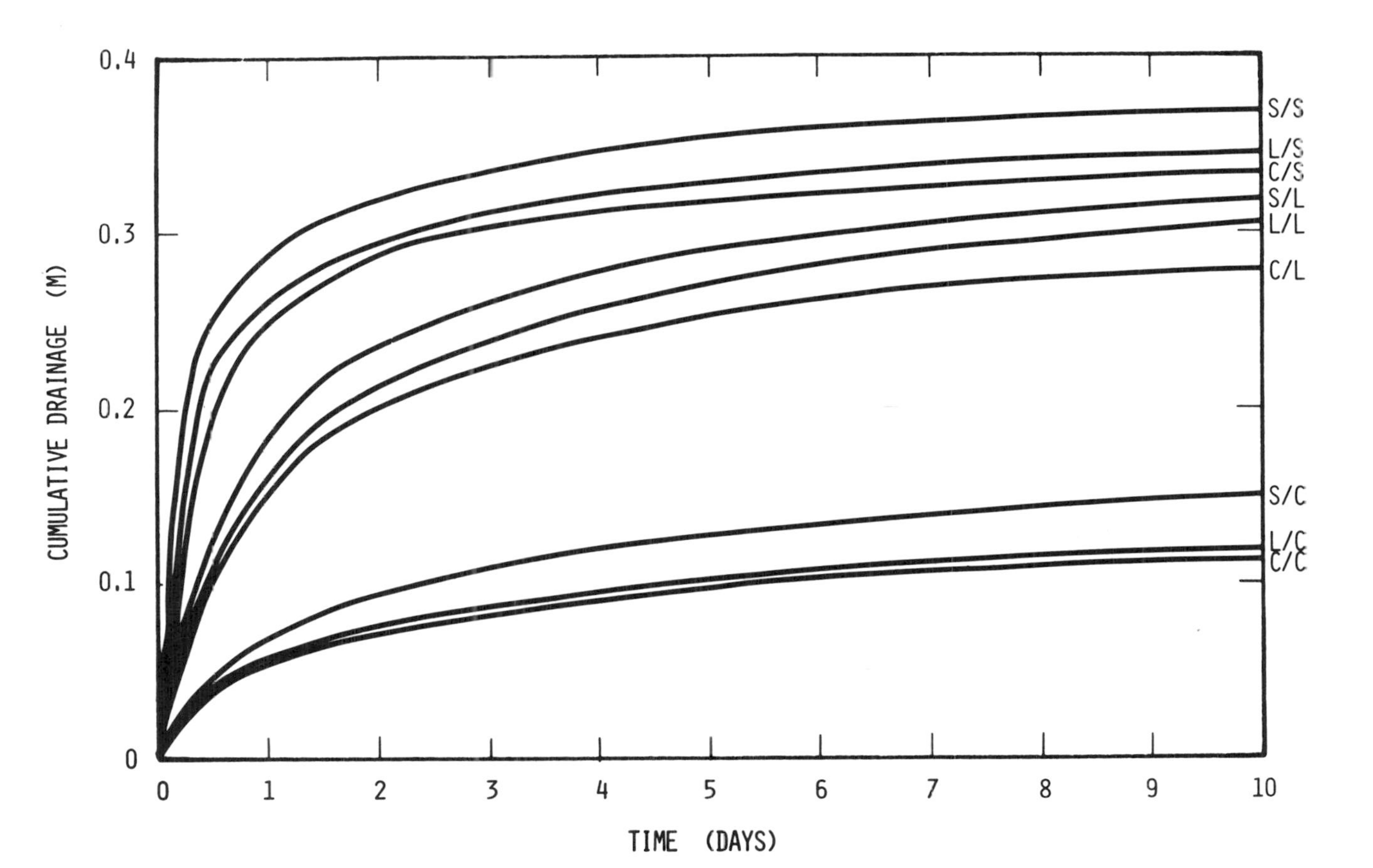

Figure 3.16. Cumulative drainage from initially saturated 2-layer profiles. S, L, C designate sand, loam, and clay, respectively.

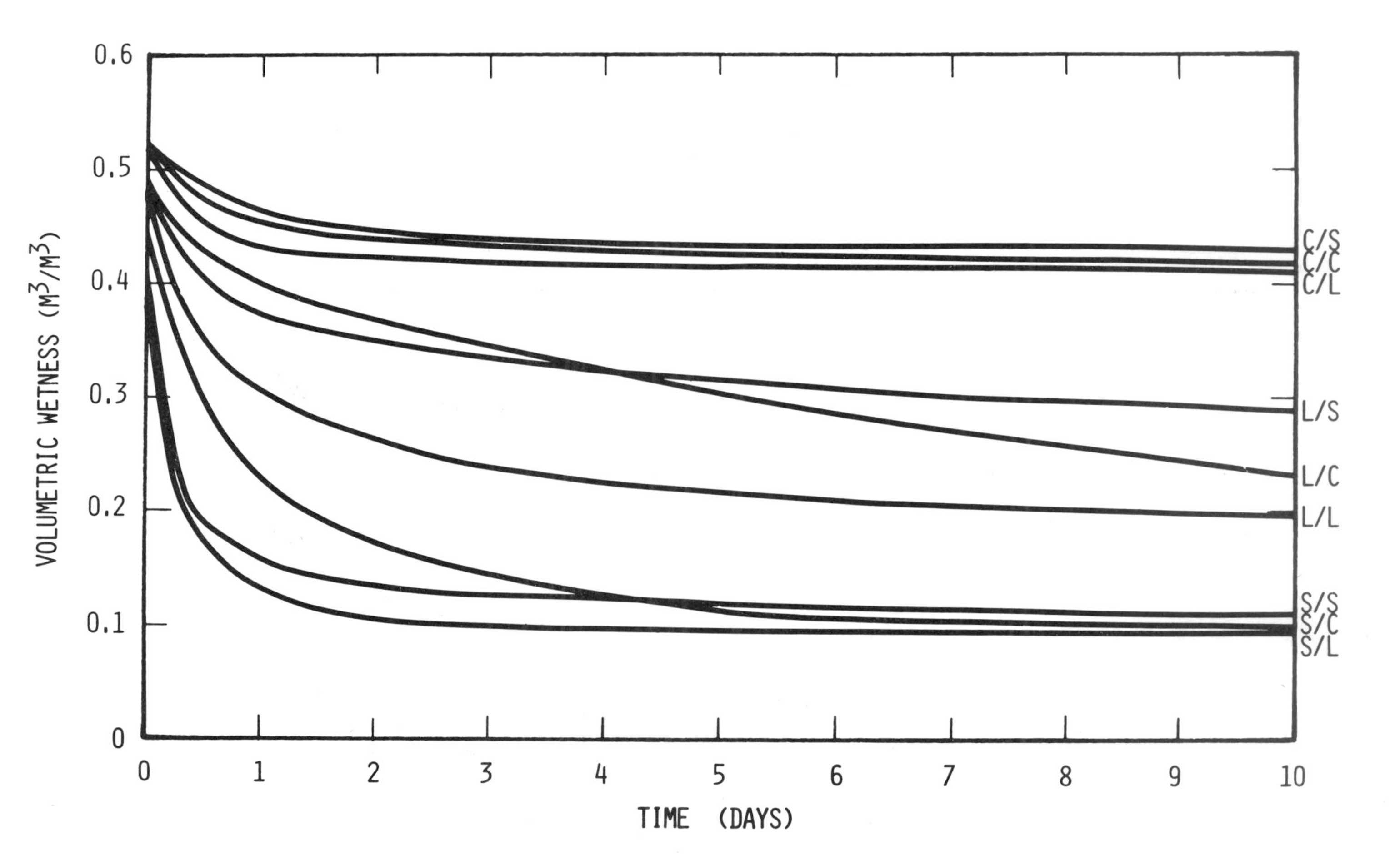

Figure 3.17. Change of volumetric wetness in the top layer (depth 0.14 m) during a drainage of initially saturated layered profiles of sand, loam, and clay.

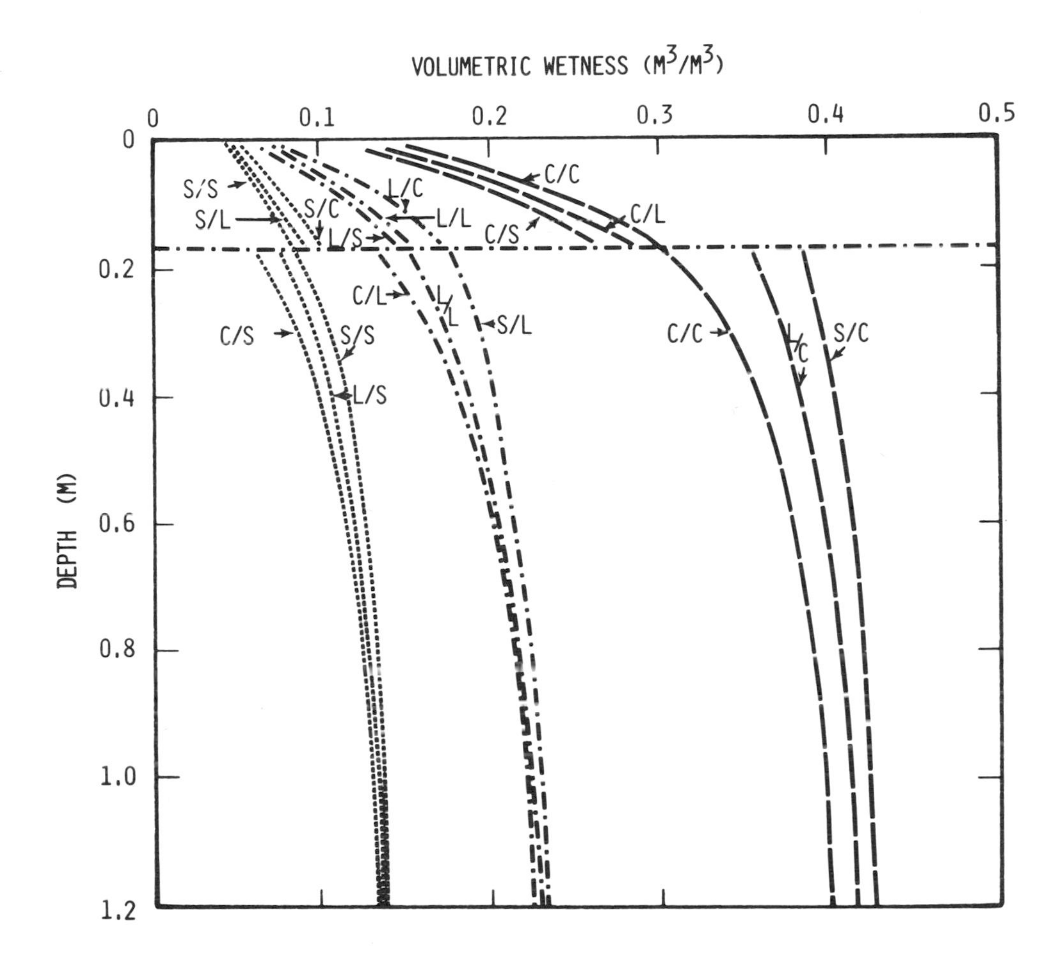

Figure 3.18. Moisture distributions in 2-layer profiles after 10 days of simultaneous drainage and evaporation.

as already mentioned, to the rapid decrease in conductivity of the sand as against the maintenance of a relatively high conductivity (and downward flux) in the clay in the range of suction which develops during drainage. The data of Figure 3.17 are pertinent to the "field capacity" concept, which is often and erroneously assumed to be an exclusive property of an individual layer (*e.g.*, its wetness at a suction of 1/3 bar) whereas in fact it is strongly influenced by the rate of drainage of the profile as a whole.

Figure 3.18 shows the distribution of moisture in two-layer profiles at the end of a 10-day period of simultaneous drainage and evaporation. Here another effect comes into play, namely the tendency of coarse-textured top layers to suppress evaporation as their conductivity falls sharply in the course of the drying process. This has the effect of preserving the wetness of the underlying fine-textured layers. An example of this effect is the wetness of the clay sublayer just under the sand top layer (depth of 0.18 m) that is 3 percent higher than under a loam top layer and fully 8 percent higher than it would be in a uniform profile of clay at the same depth. The same pattern, though to a lesser degree, can be observed for loam and sand under the various top layers.

Figure 3.19 shows the cumulative drainage from two-layer profiles during 10 days of drainage and evaporation. A comparison with Figure 3.16 shows that drainage is little affected by the evaporation process during the first five days or so. Thereafter, drainage is slowed, albeit to a small degree, in soils with a top layer of clay, owing to the persistence of evaporation.

Figures 3.20 and 3.21 are analogous to the previous two, except that they represent three-layer profiles. It is noteworthy that water retention is greatest in a layer of clay sandwiched between a top layer and a sublayer of sand, which have the dual effect of suppressing both drainage and evaporation from the intermediate layer. The occluded clay layer thus ended up after 10 days with a wetness value 5 percent higher than the corresponding depth in a uniform clay profile.

Cumulative evaporation from the two-layer and three-layer profiles during 10 days of simultaneous drainage and evaporation is shown in Figures 3.22 and 3.23. Once again, it is evident that, for each profile combination, the least evaporation occurred with a top layer of sand and the most evaporation with a top layer of clay. The highest cumulative evaporation occurred from the profile with the clay-loam-sand sequence, where the soil's ability to deliver water to the evaporating surface apparently remained highest during the 10-day period of the simulation. Had the process been continued longer, it is likely that cumulative evaporation from the uniform clay profile would eventually surpass that of all the other profiles.

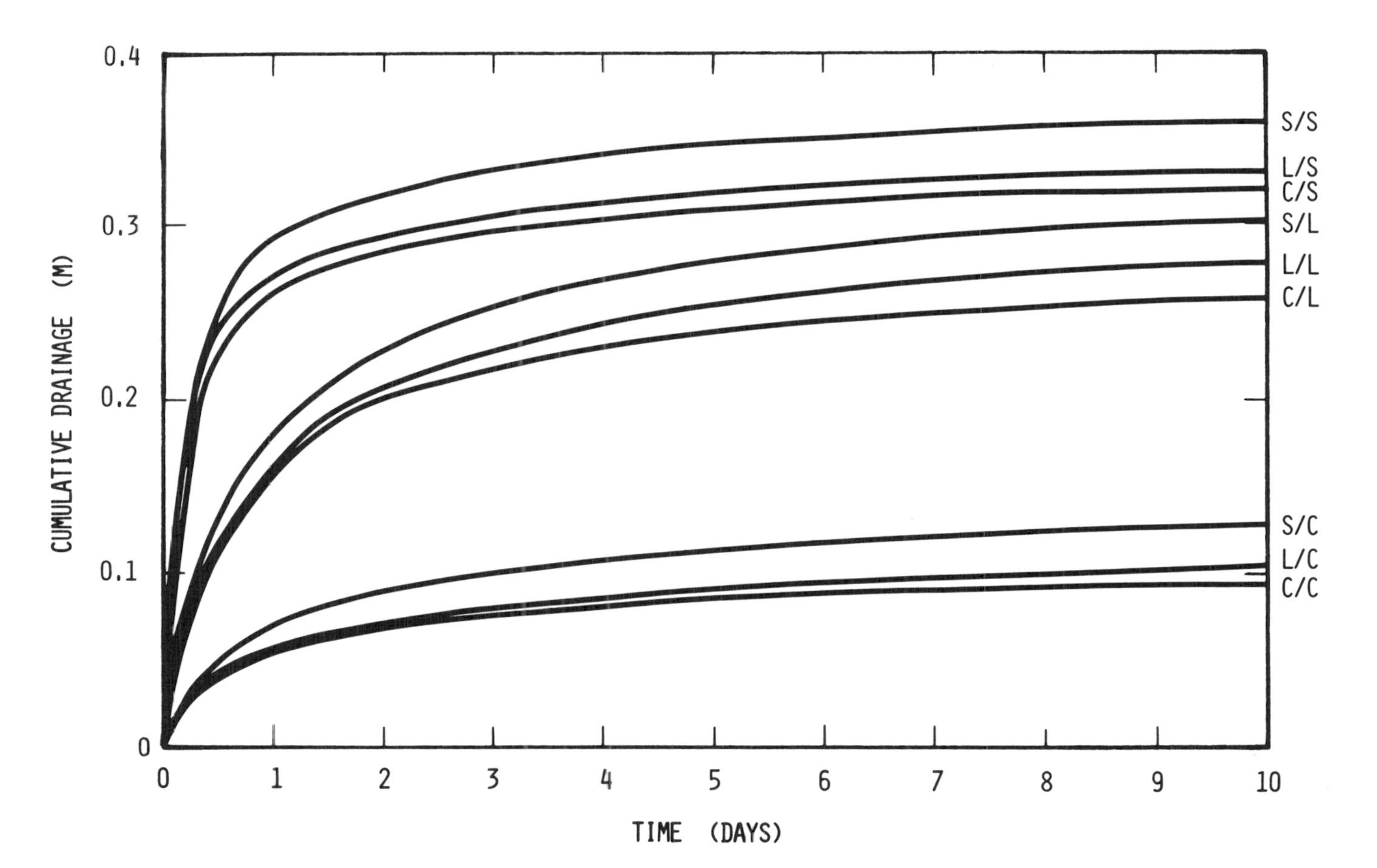

Figure 3.19. Cumulative drainage from initially saturated layered profiles during simultaneous drainage and evaporation.

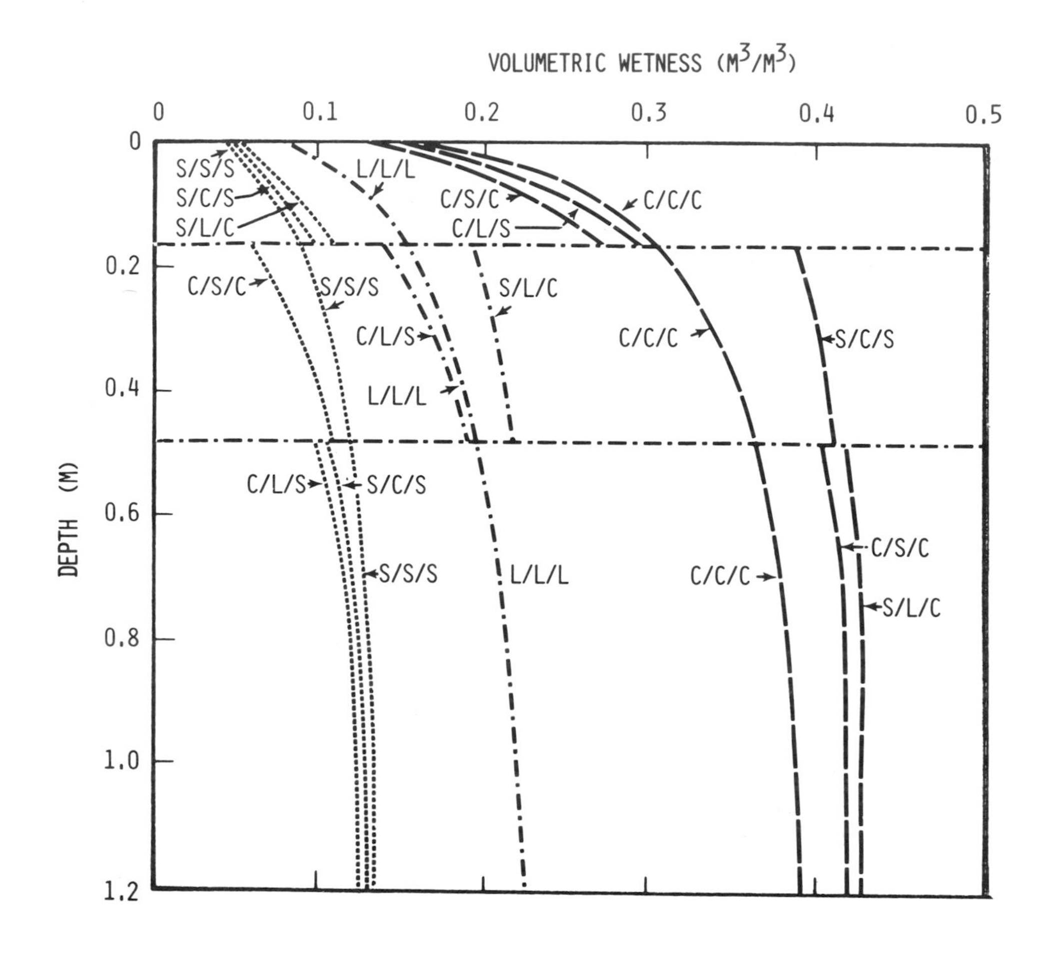

Figure 3.20. Moisture distributions in 3-layer profiles after 10 days of simultaneous drainage and evaporation.

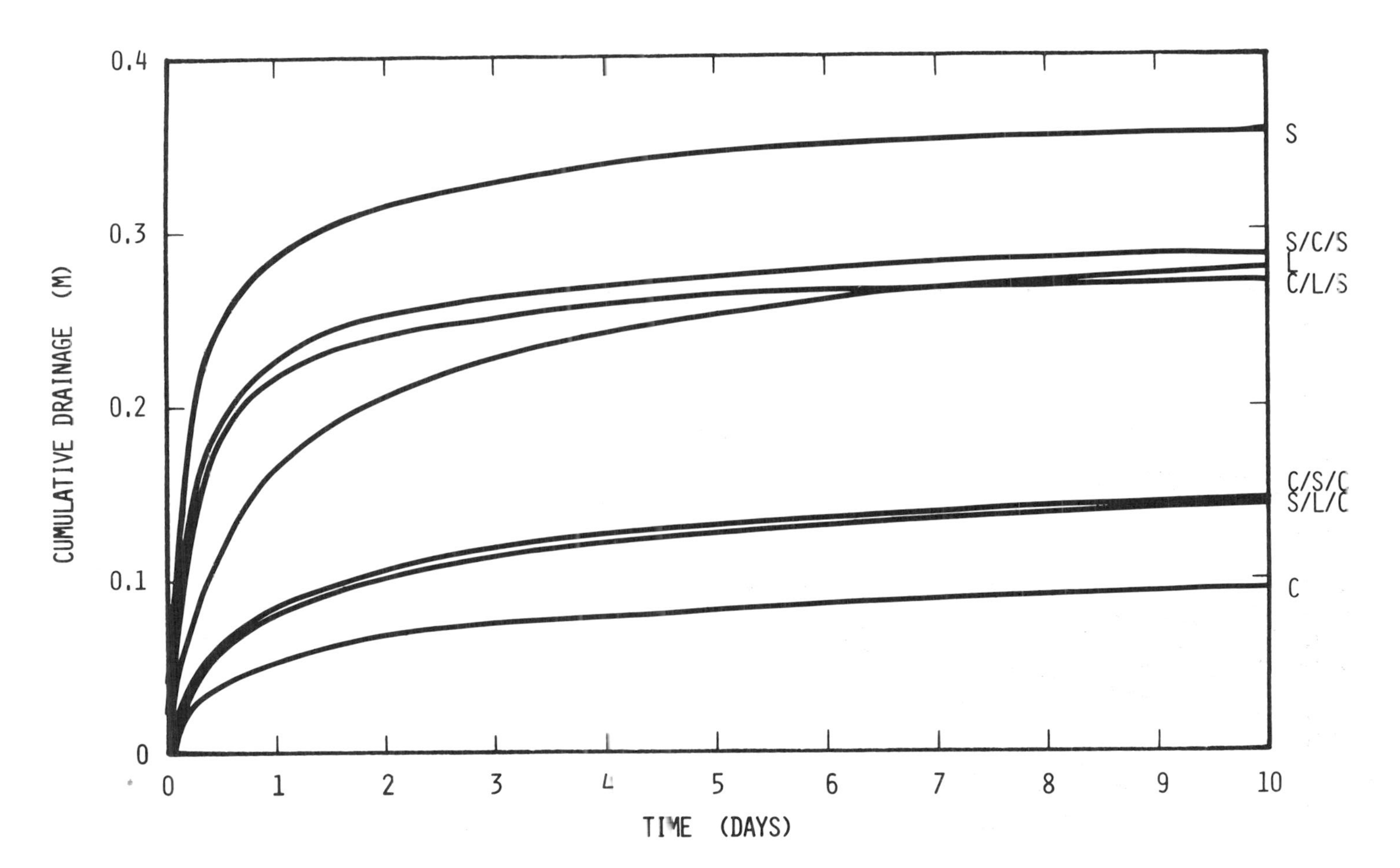

Figure 3.21. Cumulative drainage from initially saturated 3-layer profiles during simultaneous drainage and evaporation.

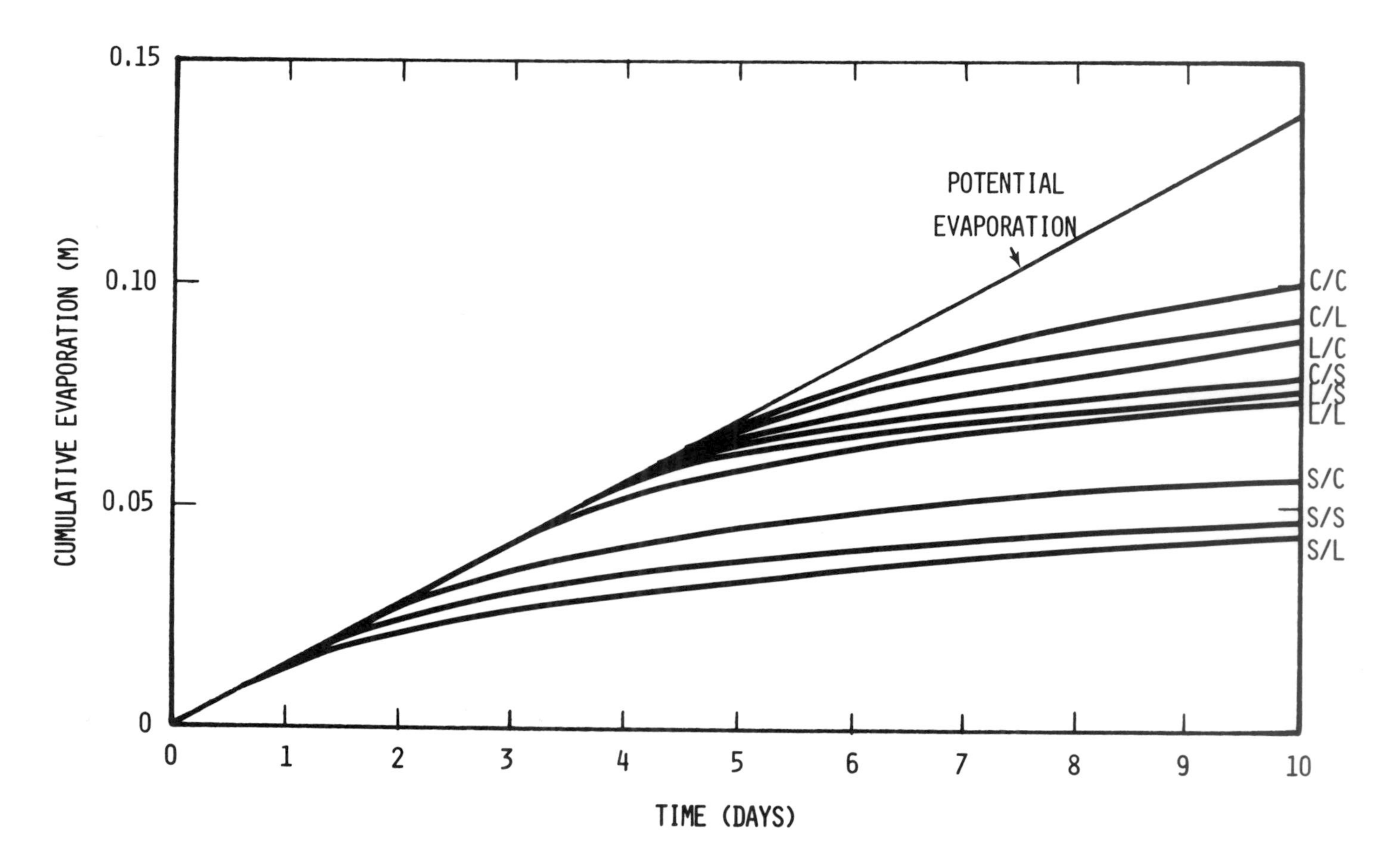

Figure 3.22. Cumulative evaporation from initially saturated 2-layer profiles during simultaneous drainage and evaporation.

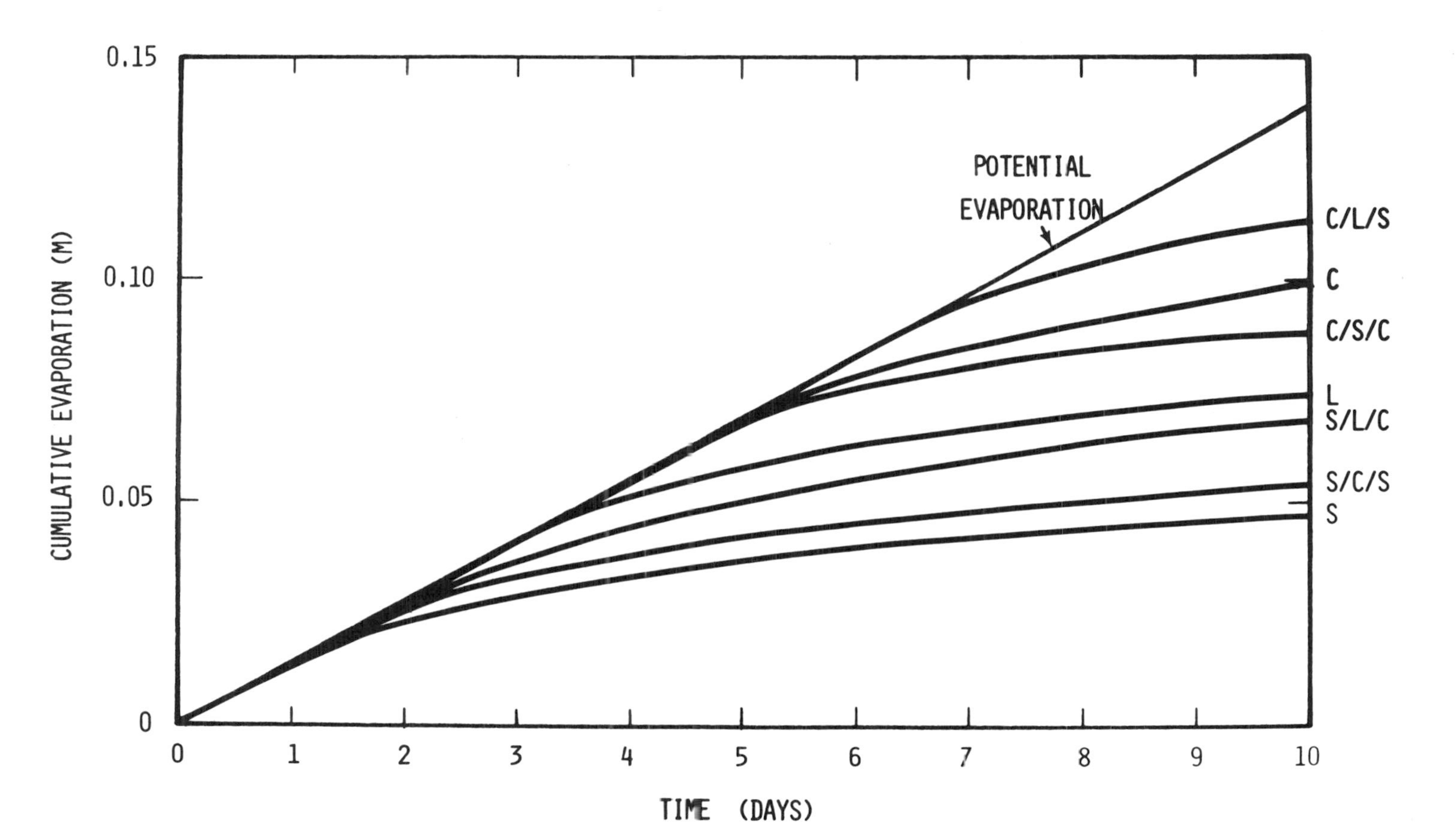

Figure 3.23. Cumulative evaporation from initially saturated 3-layer and uniform profiles during simultaneous drainage and evaporation.

Figure 3.24 indicates the change in soil wetness near the bottom of the top layer (depth of 0.14 m) in each of the two-layer profiles during 10 days of simultaneous drainage and evaporation. A comparison with Figure 3.17 shows the same clustering of curves into three groups, representing profiles with top layers of sand, loam, and clay. However, the same comparison shows quantitatively that the evaporation process strongly reduced the wetness of the clay and loam top layers in comparison with the drainage process acting alone.

Figure 3.25 shows the change in soil wetness near the top of the sublayer (at a depth of 0.18 m). In each case, the sublayer had retained the highest wetness value when covered by a top layer of sand, and the lowest wetness when covered by clay.

Figures 3.26 and 3.27 represent the profile moisture distributions during the steady-state phase of infiltration into two-layer and three-layer profiles, respectively. As is clearly evident, the presence of a fine-textured top layer impedes infiltration into underlying coarse-textured layers so that the latter are prevented from attaining saturation (as they would if the fine-textured top layer did not exist). This accords with the description of steady infiltration into crust-capped profiles given by Takagi (1960) and by Hillel and Gardner (1969). The oft-repeated statement that water does not move from a fine to a coarse layer until the latter is completely saturated is not universally true, as the pattern of flow from layer to layer depends on the relative hydraulic conductivity at the prevailing suction.

Figure 3.28 shows the cumulative infiltration into the various layered profiles in comparison with that into uniform profiles of sand, loam, and clay. In each case, the processes eventually attained a steady state (undoubtedly affected by our selection of a bottom boundary condition), but the rate of steady infiltration, as well as the pattern of transient infiltration leading up to it, differs widely among the profiles, as one or another of the layers can limit the rate of infiltration at various stages of the process.

We come finally to an overall summary of the water storage efficiencies of the variously composed profiles under a simulated weather regime consisting of an initial rainstorm followed by two days of redistribution and evaporation followed by a second rainstorm identical to the first and an eight-day period of redistribution and evaporation. The data are summarized in Table 3.2. With the two rainstorms totaling 144.4 mm, with a maximum intensity of 18 mm/hr, there was complete infiltration in all of the profiles. Doubling the maximal rain intensity without changing its amount did, however, cause some runoff from the clay-topped profiles. In any case, the amount of rainfall applied was still insufficient to cause any appreciable through-drainage. Hence

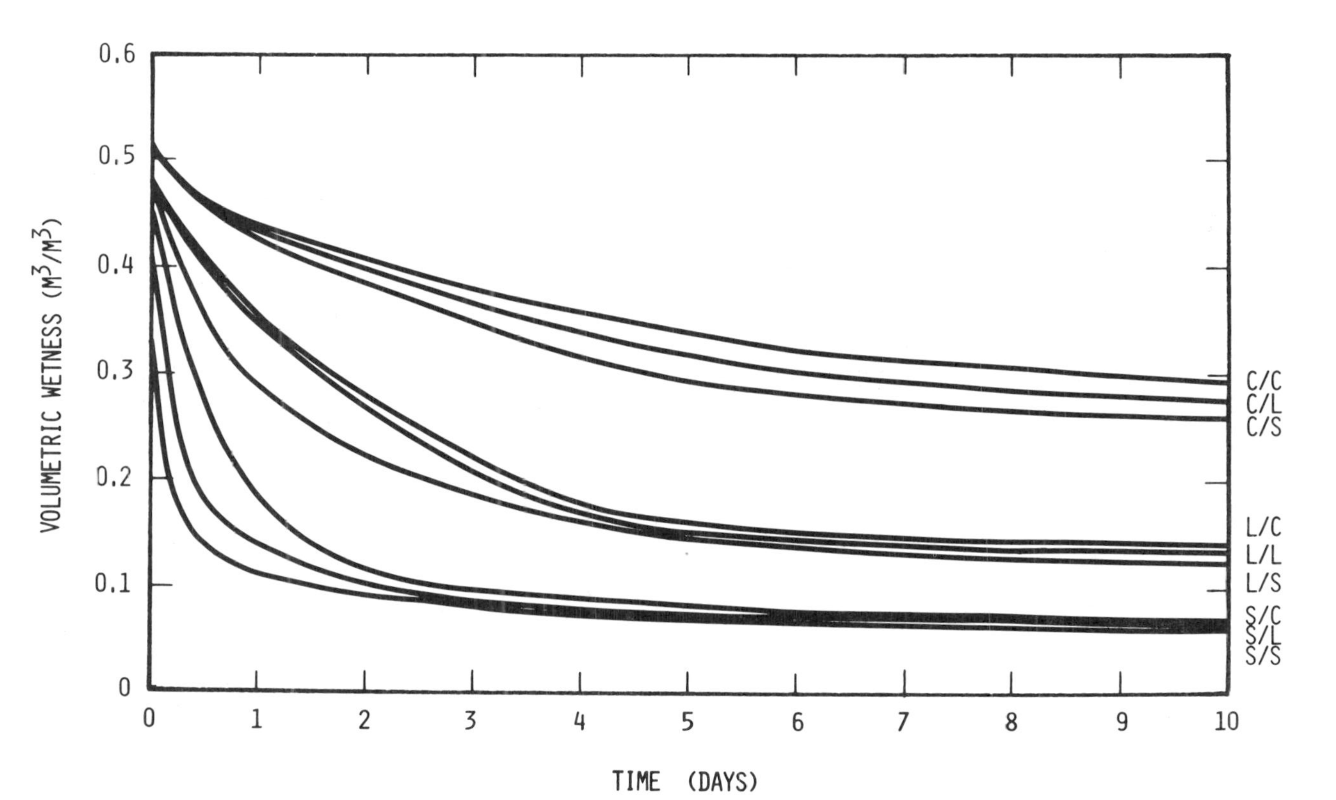

Figure 3.24. Change of volumetric wetness in the top layer (depth 0.14 m) during simultaneous drainage and evaporation from initially saturated 2-layer profiles.

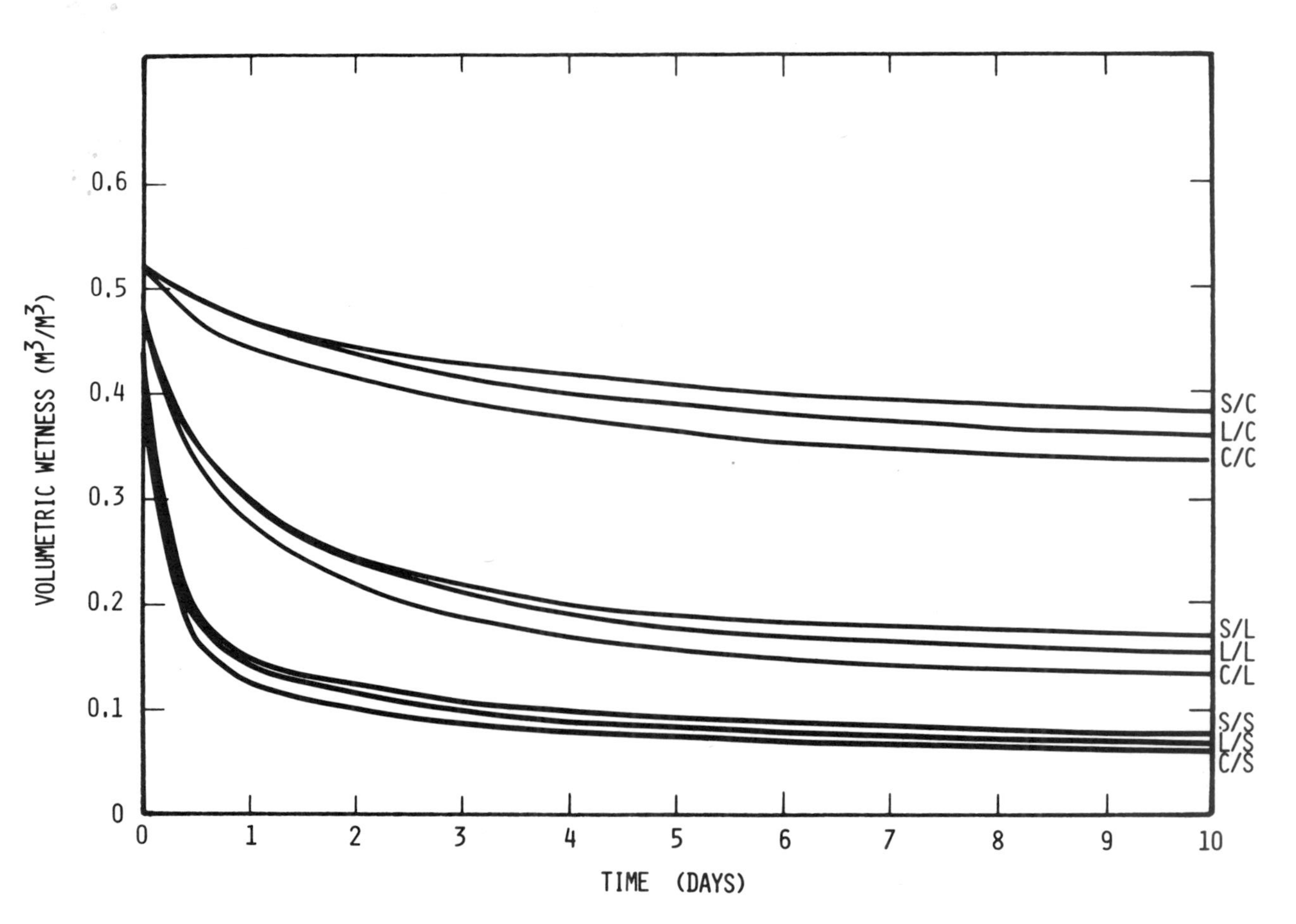

Figure 3.25. Change of volumetric wetness in the sub layer (depth 0.18 m) of 2-layer profiles during simultaneous drainage and evaporation.

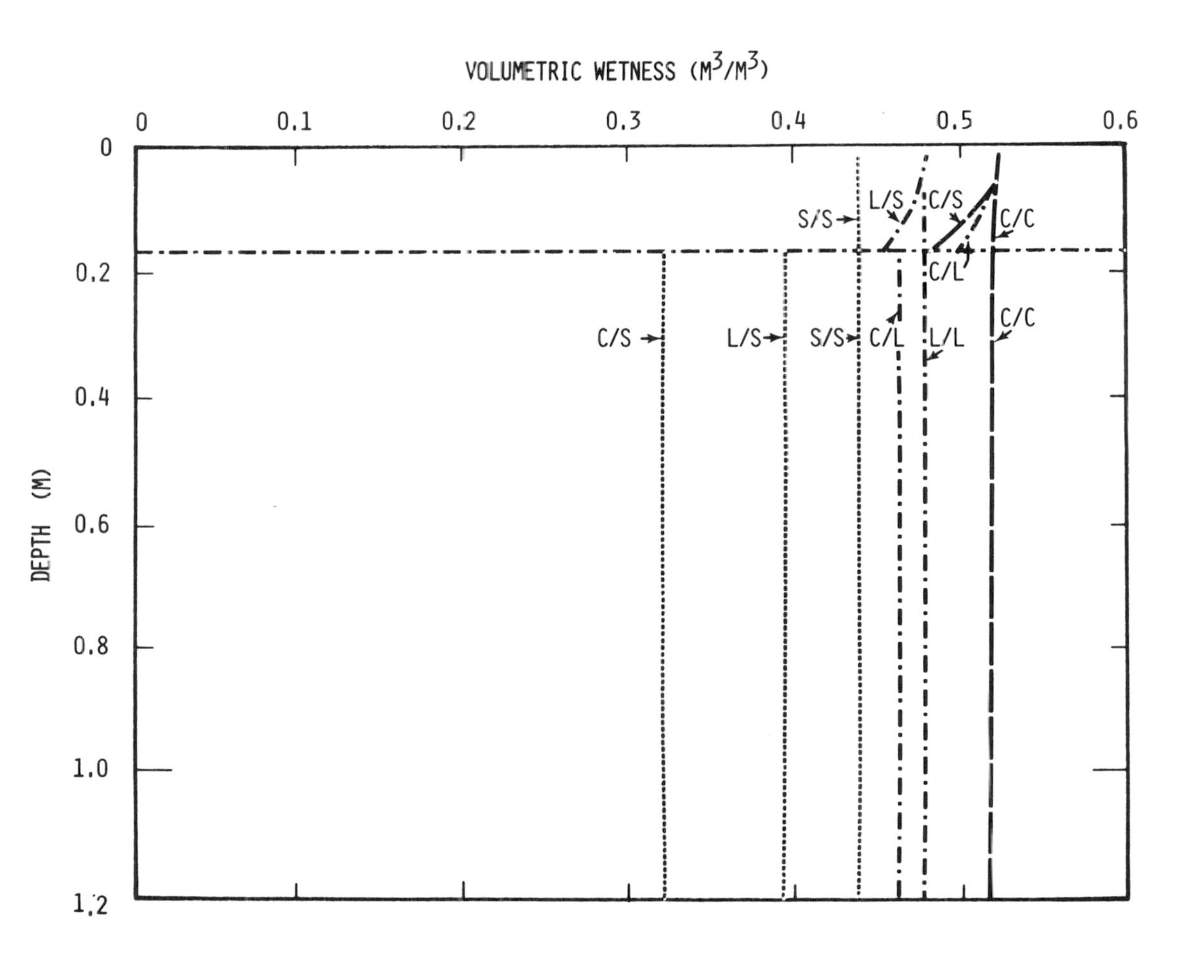

Figure 3.26. Moisture distributions in 2-layer profiles during steady infiltration.

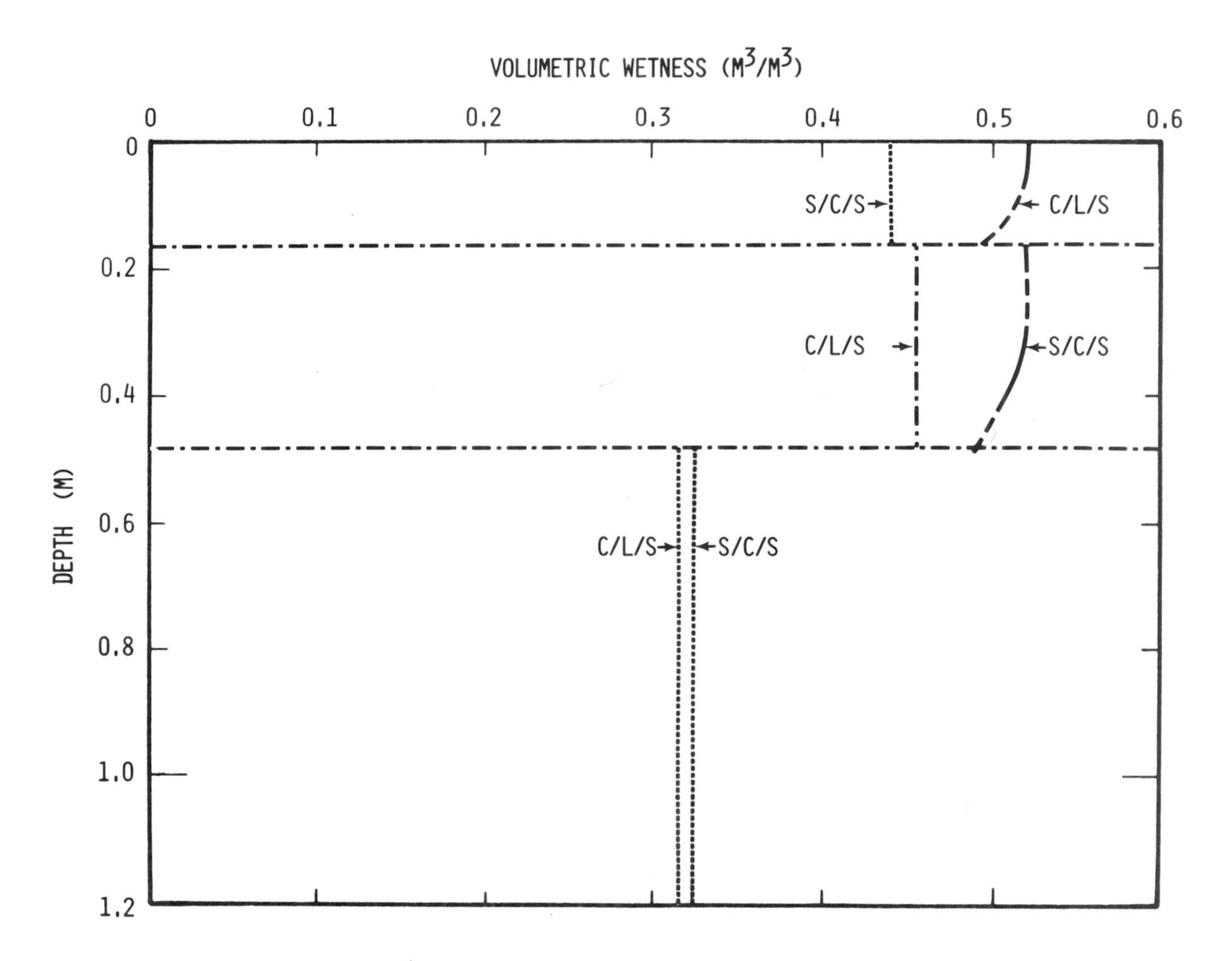

Figure 3.27. Moisture distributions in 3-layer profiles during steady infiltration.

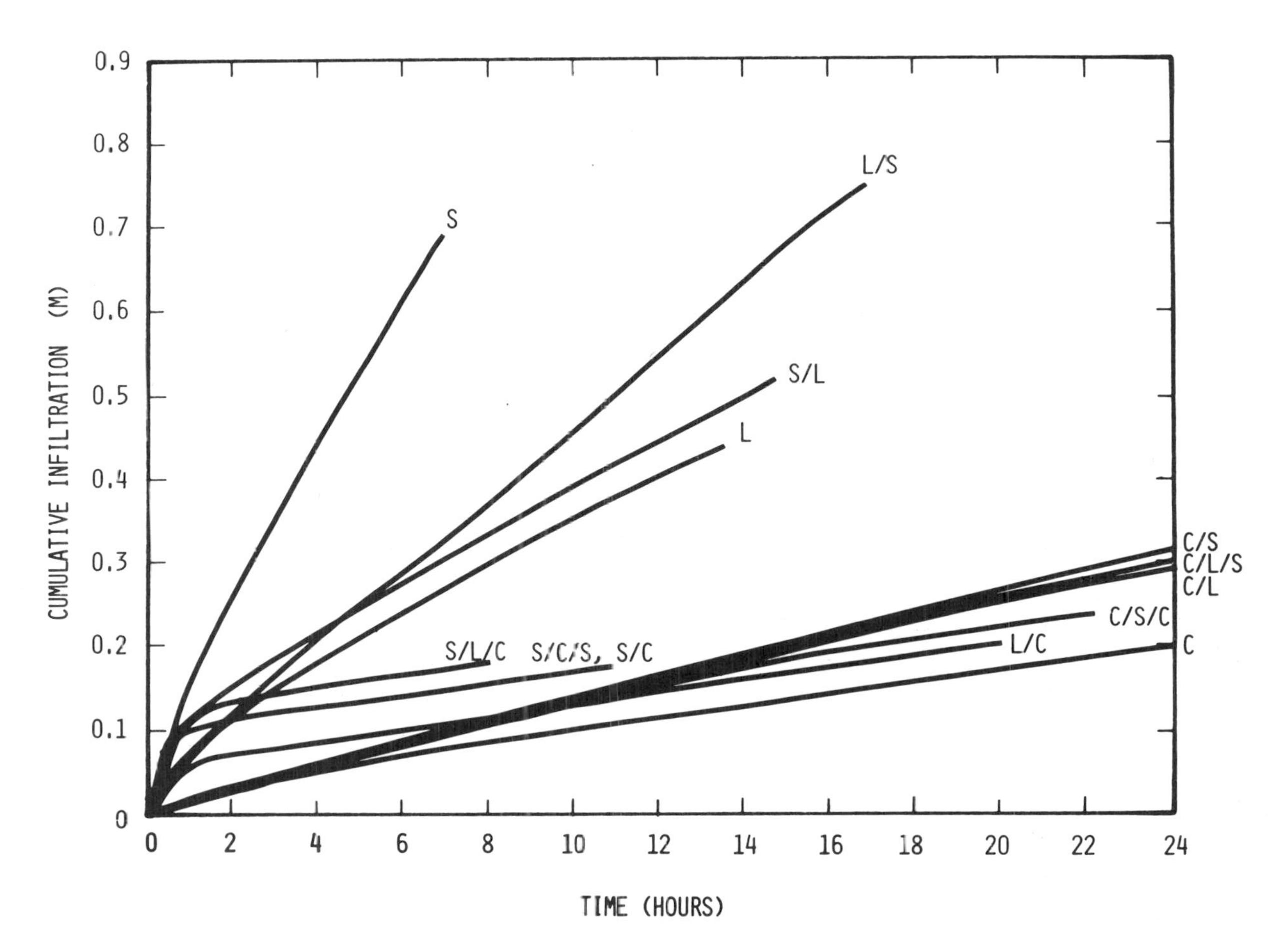

Figure 3.28. Cumulative infiltration into layered and uniform profiles of sand, loam, and clay.

Table 3.2. Water balance of layered profiles for a 10-day simulation, including two rainstorms (first and third nights) totaling 144.4 mm, and two periods of evaporation (the first during the two days between rainstorms; the second during the eight days after the last rainstorm). Symbols S, L, and C stand for sand, loam and clay. S/C represents a two-layer profile of sand over clay, and other layered profiles are designated accordingly. Profiles with asterisk were subjected to the same rainfall with doubled intensity.

Soil Profile	Total Infiltration (mm)	Total Runoff (mm)	Total Evaporation (mm)	Total Drainage (mm)	Storage Increment (mm)	Storage Efficiency (%)
Sand	144.4	0.0	66.3	1.2	76.9	53.3
Loam	144.4	0.0	78.7	0.0	65.7	45.5
Clay	144.4	0.0	87.3	0.0	57.1	39.5
S/L	144.4	0.0	51.2	0.5	92.7	64.2
S/C	144.4	0.0	49.8	0.02	94.6	65.5
L/S	144.4	0.0	93.4	0.0	51.0	35.3
L/C	144.4	0.0	70.7	0.02	73.7	51.0
C/S	144.4	0.0	97.4	0.02	47.0	32.5
C/L	144.4	0.0	94.4	0.01	50.0	34.6
S/L/C	144.4	0.0	47.5	0.03	96.9	67.1
C/L/S	144.4	0.0	104.7	0.0	39.7	27.5
C/S/C	144.4	0.0	89.6	0.02	54.8	38.0
S/C/S	144.4	0.0	60.9	0.0	83.4	57.8
C/S/L	144.4	0.0	90.3	0.04	54.1	37.5
L/S/C	144.4	0.0	85.3	0.02	59.1	40.9
C/S*	119.1	25.3	91.4	0.0	28.7	19.9
C/L/S*	122.1	22.3	92.7	0.0	29.4	20.4

the major differences in water balance among the various profiles were due to evaporation.

We have already defined *storage efficiency* as the ratio of the total amount of water added to the profile from beginning to end of the 10-day simulation period to the total rainfall. We find that the highest storage efficiency values were exhibited by profiles with coarse-textured upper layers and fine-textured bottom layers. The lowest profile storage efficiencies occurred in profiles with the opposite sequence of layers, *i.e.*, fine-textured over coarse-textured. It should be stressed, however, that the actual and relative values of storage efficiency in any particular case not included in our necessarily limited set of examples would depend not only upon profile constitution but also upon the weather pattern and the period of time considered. Our choice of an example in which evaporation, rather than runoff or drainage, is the dominant process of water loss is thought to be pertinent to the conditions which might prevail in some semi-arid areas. Where rain intensities are high, runoff can be expected to constitute an important component of the water balance of clay-topped profiles, whereas drainage can be important in areas of high total rainfall and predominantly sandy soils. The least effective storage from the combined standpoints of evaporation, drainage, and runoff can be expected in profiles with a shallow top layer of clay over sand.

Discussion

We wish to re-emphasize the theoretical nature of this study. In particular, the examples chosen to illustrate the model and its possible consequences were based on arbitrary soil properties and weather patterns. The soil hydraulic characteristics which we assigned to our so-called "sand," "loam," and "clay" may or may not be typical of soils which are generally described by those same terms. Such is the range and extent of soil diversity that it is doubtful if such a thing as a "typical" soil of any textural class can be defined. A further qualification is necessary in connection with the calculation of the hydraulic conductivity function. While the same procedure, based on the capillary hypothesis, was used for all three soil types, we are aware that in principle this hypothesis is more applicable to sand than to clay.

Our model differs from earlier ones (*e.g.*, Stroosnijder *et al.* 1972) in that it provides a more comprehensive description of the sequential or simultaneous processes of infiltration, redistribution, and evaporation in variously constituted profiles. The numerical simulation approach is seen to be most versatile, as it is unencumbered by the restrictive assumptions necessary for obtaining analytical solutions (*e.g.*, Aylor and Parlange 1973). Specifically, our model can handle flows in both upward and downward directions for profiles of both uniform and highly non-uniform initial conditions in which soil hydraulic properties may vary in depth either gradually or abruptly from layer to layer. Furthermore, the model can accommodate rain or evaporation episodes of varying duration and intensity. On the other hand, our model fails to take into account a whole array of interactions which will undoubtedly call for much further study. First of all, our model is one-dimensional (vertical), while heterogeneous soils in the field seldom behave one-dimensionally. Second, there is the hysteresis phenomenon, which comes into play during cyclic processes such as redistribution (Rubin 1967) and diurnally fluctuating evaporation (Hillel 1976). The hydraulic behaviour of the soil, particularly in the all-important surface zone, can vary in time, as under raindrop impact and wetting-drying cycles. Moreover, temperature and solute effects can also have a bearing on soil hydraulic behaviour, as can soil moisture extraction by roots (Hillel *et al.* 1975c). Finally, our model disregards the problem of wetting-front instability which can be important in some cases (Raats 1973; Philip 1975).

The principal conclusion of this modeling experiment is that soil water storage efficiency is strongly affected by profile layering, and can vary within the wide limits of 20% to 60% in a profile composed of layers of sand, loam, and clay.

IV. HYDROLOGY OF A SLOPING FIELD, INCLUDING SURFACE RUNOFF AND GROUNDWATER FLOW

A. Description of the Problem

One of the most vital and challenging frontiers of research at the present time is the no-man's land between the heretofore separate disciplines of soil physics and hydrology. Soil physicists have traditionally dealt with phenomena on the scale of a soil profile, most often assumed to be a one-dimensional (vertical) system of the sort we have depicted in our first three chapters. Hydrologists, on the other hand, have attempted to deal with soil-water phenomena on the scale of a complete, three-dimensional and generally heterogeneous *watershed* or *drainage basin.* To deal directly with such a complex system, hydrologists have had to devise empirical and statistical, rather than mechanistic, approaches.

Traditionally, the science of hydrology has consisted of two main branches: *surface-water hydrology,* dealing with water flowing or stored over the soil surface or in streams and lakes; and *ground water hydrology,* dealing with flow and storage below the water table. In between the two realms lies the so-called *unsaturated zone* of the soil, a realm too-often neglected by hydrologists but of central interest to soil scientists. Nowadays, however, hydrologists have universally come to realize the importance of the unsaturated zone as a primary determinant of runoff formation and quantity, as well as of subsurface water flow paths and velocities. A better understanding of soil-water dynamics is also essential in connection with the movement of dissolved pollutants and the progress of various chemical and biological changes to which they are subject as they pass through the soil (Amerman 1973).

Soil physicists have, in the last two decades, achieved some considerable progress in the mechanistic formulation of the principles of water movement in unsaturated as well as saturated regions of the soil, and therefore ought to be able to provide hydrologists with knowledge concerning the missing link in hydrology even while expanding their own horizons as environmental scientists. The problem is, however, how to re-compose the complex hydrological whole from the sum of its soil-physical parts. A simple extrapolation of one-dimensional processes does not yield a complex three-dimensional system unless such additional phenomena as lateral heterogeneity and overland as well as groundwater flow are also formulated and taken into consideration on a mechanistic basis. Mathematical simulation can be a valuable, perhaps even indispensable, tool in any attempt to incorporate soil-physical phenomena into a comprehensive, physically based treatment of composite hydrological systems.

Perhaps the most plausible place to attempt a reconciliation between soil physics and hydrology is in modeling the dynamics of water in an agricultural field. Consider a sloping field, uniform along its contours (the y-axis) but not necessarily uniform in depth (z-axis) or slope-direction (x-axis). Such a field can be represented in terms of a two-dimensional profile (Figure 4.1) or cross-section, being an extension of the usual one-dimensional representation of a soil profile. Modeling the vertical and horizontal components of water-flow processes in such a system is indeed an intermediate stage in the application of soil physics to watershed hydrology. Herein, we propose to present a simple version of such a model, capable of portraying the processes of infiltration, surface storage and runoff, evaporation, internal unsaturated soil moisture movement, as well as groundwater recharge and discharge, as shown schematically in Figure 4.1.

B. Governing Equations

Overland Flow

The basic equation governing the rate of flow (discharge) in a channel is

$$Q = VA \tag{4.1}$$

where Q is the flow rate, V average velocity, and A cross-sectional area of flow.

Conservation of matter ("continuity") requires that, in the absence of rainfall and infiltration, the change in water level h with time be equal to the negative of the change of flow rate with distance, *i.e.*,

$$\frac{\partial h}{\partial t} = - \frac{\partial Q}{\partial x} \tag{4.2}$$

If rain and infiltration are taken into account, we have

$$\frac{\partial h}{\partial t} + \frac{\partial Q}{\partial x} = r - i \tag{4.3}$$

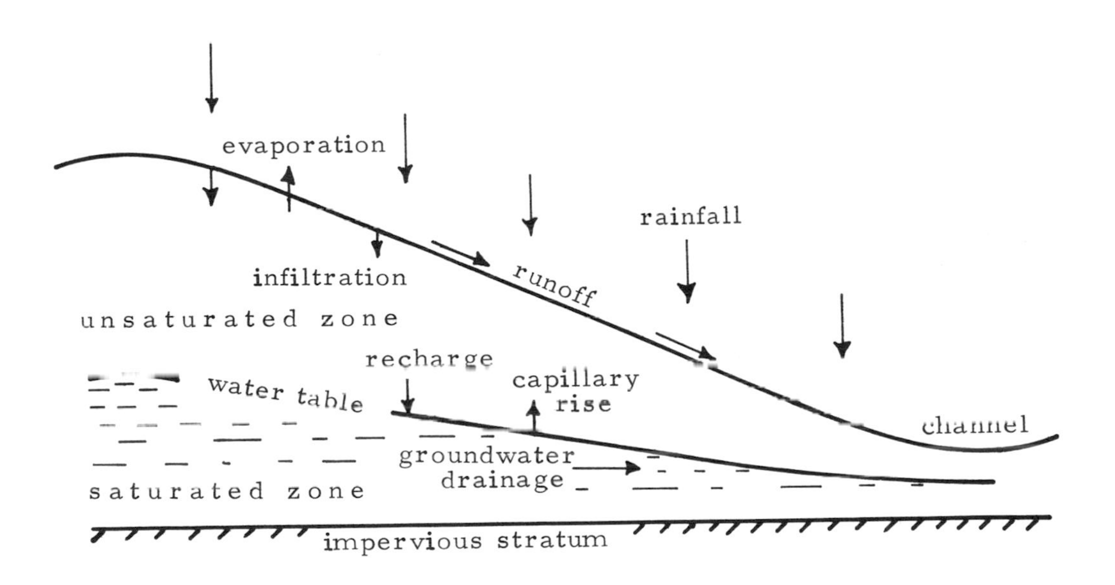

Figure 4.1. Schematic depiction of the hydrology of a sloping field. The two-dimensional cross-sectional view portrays the profile of a field assumed to be homogeneous along its contour lines (perpendicular to the plane of the drawing).

Mean velocity and cross-sectional area of flow depend on shape and size of channel. Since flow is inhibited by shear stresses resulting from the immobility of the water in immediate contact with the sides and bottom of the channel, the hydraulic resistance of a channel is generally proportional to the area of contact per unit volume of flowing liquid (or to the length of the perimeter of contact per units cross-sectional area of flow). This contact area can be characterized in terms of the *hydraulic radius,* or *hydraulic mean depth,* of a channel, defined as the cross-sectional area divided by the wetted perimeter.

Now consider an infinitely wide channel without sides. Such a "wide-open channel" is represented by a laterally uniform sloping surface with a layer of water flowing over it, as illustrated in Figure 4.2. Here the effective "hydraulic radius" is h. As an approximation, let us assume that the two forces acting on the flowing water are: (1) the component of the water's weight acting in the direction of the bed slope, and (2) the shear stresses developed at the solid-to-water boundary. If the velocity is more-or-less uniform along the slope, then the two forces must be approximately balanced, and we get:

$$\tau_o = \rho g h \alpha \tag{4.4}$$

where ρ is the liquid's density, g the acceleration of gravity, h the water depth, α the slope, and τ_o the bottom shear stress.[1]

Shear stress τ_o is known to be related to the average velocity squared:

$$\tau_o = a\rho V^2 \tag{4.5}$$

where a is a proportionality factor. Combining the last two equations gives:

$$V = (g/a)^{\frac{1}{2}} h^{\frac{1}{2}} \alpha^{\frac{1}{2}} = C h^{\frac{1}{2}} \alpha^{\frac{1}{2}} \tag{4.6}$$

---

[1]A more rigorous approach is to start from the momentum equation:

$$\frac{\partial V}{\partial t} + \frac{\partial V}{\partial x} + g\frac{\partial h}{\partial x} = [r - f]\frac{V}{h} - \frac{\tau_o}{\rho h} + g\alpha$$

Here the underlying assumptions are that the bottom slope is small, the velocity distribution uniform, and the overpressure from rainfall is negligible. If we also assume that rainfall and infiltration have little effect on flow dynamics and that the slope of the free surface ($g\ \partial h/\partial x$) and the inertial terms ($\partial V/\partial t + V\ \partial V/\partial x$) are small compared to friction ($\tau_o/\rho h$) and bottom slope ($g\alpha$), we arrive at: $\tau_o = \rho g h \alpha$).

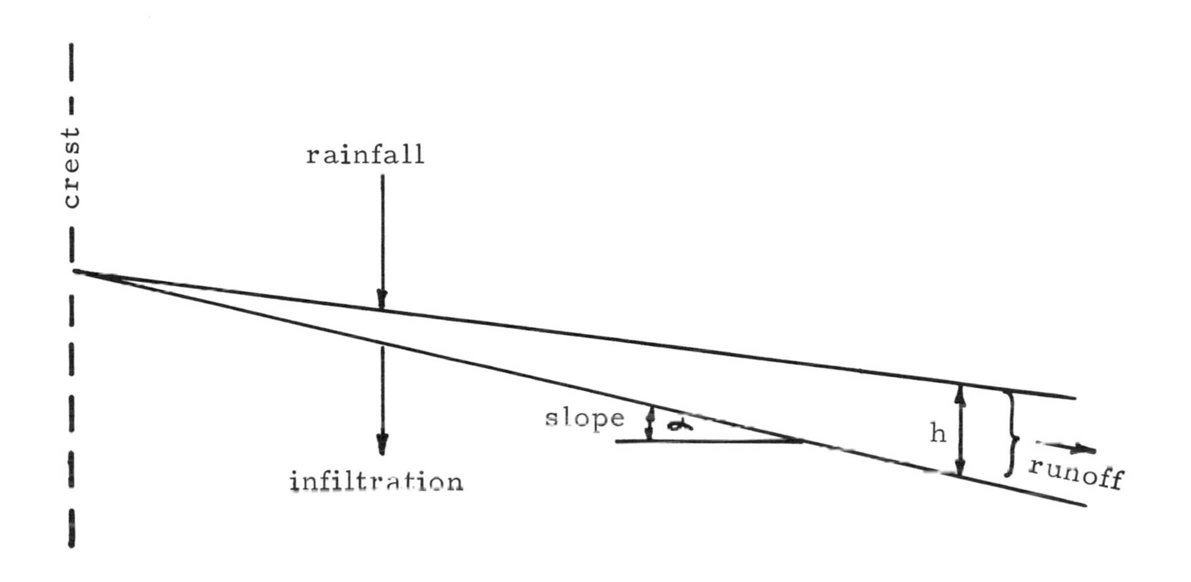

Figure 4.2. Surface-water excess and sheet overland flow (schematic).

where C is the *Chezy coefficient,* named after the French engineer who first discovered this relationship some two hundred years ago. A large amount of empirical work conducted since has indicated that Chezy's coefficient C is not a constant but depends on water depth h and surface roughness n as follows:

$$C = \frac{h^{1/6}}{n} \tag{4.7}$$

Combining this with equation (4.6) gives the useful equation generally attributed to Robert Manning (1891):

$$V = h^{2/3} \alpha^{1/2}/n \tag{4.8}$$

Multiplying by h generally converts the left side of this equation to discharge per unit width:

$$Q = h^{5/3}\alpha^{1/2}/n \approx h^2\alpha^{1/2}/n \tag{4.9}$$

In practice, a power of 2 for h rather than 5/3 has been found to give satisfactory results for overland flow. Measured values of the roughness coefficient have been tabulated (*e.g.* Sellin 1969), and range between 0.01 for extremely smooth channels to 0.1 for flood plains with a growth of heavy timber. A value of about 0.03 seems to be reasonable for soil surfaces (either bare or with a cover of short grass).

The last version of Manning's equation can be combined with equation (4.3) to yield the *kinematic wave equation:*

$$\frac{\partial h}{\partial t} + \beta h^{m-1}\, \partial h/\partial x = r - f \tag{4.10}$$

where $\beta = \alpha^{1/2}/n$, and $m = 2$ for equation (4.9).

Groundwater Flow

Below the water-table the pressure of soil water is greater than atmospheric, the soil is saturated, and flow is governed by the saturated hydraulic conductivity. Darcy's law alone is sufficient to describe steady flow processes, but for unsteady flow, Darcy's law must be combined with the mass-conservation law to obtain the *Laplace equation* (Hillel 1971):

$$\nabla^2 H = 0 \tag{4.11}$$

The direct analytical solution of Laplace's equation for conditions pertinent to groundwater flow is not generally possible. Therefore, it is often necessary to resort to approximate or indirect methods of analysis. In the solution of problems relating to flow toward a shallow sink (*e.g.*, a drainage channel), it is often convenient to employ the so-called *Dupuit-Forchheimer assumptions* (Forchheimer 1930). These assumptions are that in a system of gravity flow toward a shallow sink, all the flow is horizontal and that the velocity at each point is proportional to the slope of the water table but independent of the depth (van Schilfgaarde 1974). Though these assumptions are obviously not correct in the strict sense and can in some cases lead to anomalous results (Muskat 1946), they often provide feasible solutions in a form simpler than obtainable by rigorous analysis. They apply most suitably to cases in which the flow region is of large horizontal extent relative to its depth.

Accordingly, taking the effective gradient to be equal to the slope of the water table (dh/dx) above each point, and allowing only horizontal flow, we get

$$Q = K_s h \frac{dh}{dx} \qquad (4.12)$$

where K_s is the hydraulic conductivity of the saturated soil and h is height of the water table above an impervious layer which is assumed to form the "floor" of the flow system.

C. The Conceptual Model

In the model which follows, water movement is conceptualized as a series of interacting one-dimensional processes. The lateral extent of the cross-section is bounded at the upslope end by the watershed divide and at the downslope end by a channel or stream. The vertical bounds of the system are the soil surface on top and an impervious plane below the soil.

To discretize this continuous system for the purpose of computer modeling, the hillside is divided into a series of vertical columns (Figure 4.3), not necessarily of uniform width or properties. The upper surface of each column receives or loses water according to the prevailing rainfall or evaporative regime. During each rainstorm, water reaching the soil surface is partitioned between infiltration and surface-water excess. The latter, in turn, is partitioned between surface storage and surface runoff (overland flow). Runoff is portrayed as a one-dimensional process directed downslope and routed over the upper surfaces of the vertically displaced successive columns. The vertical displacements of successive columns can be made equal or variable, depending upon uniformity of the slope.

Infiltration is calculated for each column as described in the preceding chapter. Water is moved vertically through a series of compartments of various thicknesses, which may also differ

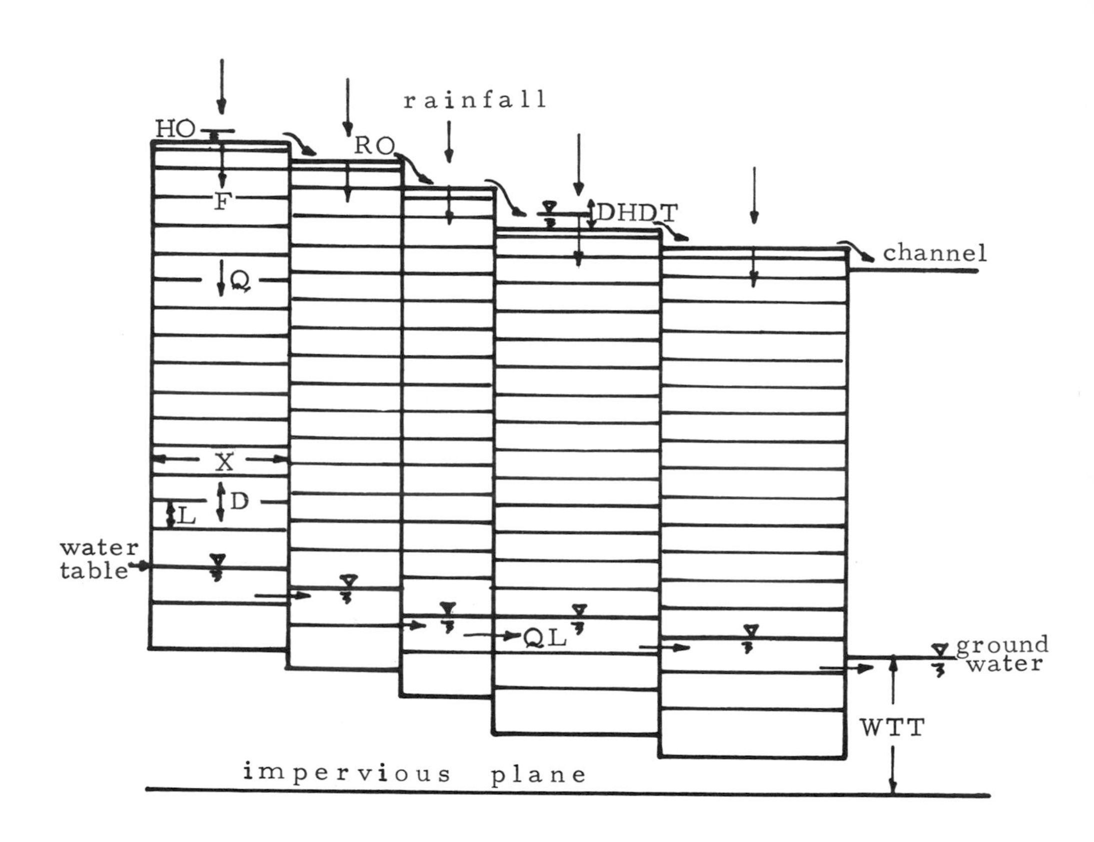

Figure 4.3. Model representation of the profile of a sloping field divided into columns and layers, not necessarily of uniform properties.

in hydraulic properties. For the sake of simplicity, the model assumes that horizontal gradients and fluxes in the unsaturated zone are negligibly small.

As the lower compartments of each soil column fill with water due to infiltration and vertical drainage, the water table rises. Saturated flow equations and the water table gradient govern the lateral flow of groundwater between adjacent columns. In this way, groundwater moves downhill in a series of one-dimensional steps. At the uphill boundary, we generally have a crest (or watershed divide) which is taken to be a zero-flow boundary. At the downhill boundary, the saturated compartments contribute water to a channel or stream which serves as a constant-level sink. The flows from these compartments and from overland runoff constitute hydrographs having units of volume per time per length of stream bank. Source and sink phenomena (*e.g.*, inflows from irrigation ditches or extraction by roots) can also be included.

All equations pertaining to unsaturated and saturated soil water flow, as well as to overland flow, are thus physically based. The solution of these equations is accomplished by discretizing the kinematic wave equation representing overland flow, the lateral groundwater equation representing flow below the water table, and the unsaturated soil moisture flow equation, as will be shown in the computer program.

D. Description of the Computer Model

The program is presented in Figure 4.4. It begins as usual with memory-space allocations (STORAGE and DIMENSION) for the variables, and EQUIVALENCE specifications for the subscripted variables to be integrated.

INITIAL Section

The following parameters or constants are assigned values:

(1) NL = number of layers (compartments) per column
(2) NC = number of columns
(3) X = width of each column
(4) IW = initial volumetric wetness
(5) SK = saturated hydraulic conductivity
(6) MINPOT = minimal matric potential of air-dry soil
(7) M = exponential constant of Manning's equation
(8) N = the surface characteristic constant of Manning's equation
(9) ALF = composite constant of Manning's equation
(10) S = slope of soil surface
(11) WTT = level of water table at downslope drain
(12) PET = potential evaporation rate.

The following are given in tabular form:

(1) TABLE L = thicknesses of compartments comprising each column's profile

(2) FUNCTION SUCTB = volumetric wetness of the soil versus matric suction head (the soil moisture characteristic, non-hysteretic)

(3) FUNCTION CONDTB = volumetric wetness of the soil versus hydraulic conductivity

(4) FUNCTION RAINTB = time versus rainfall rate.

The following values are computed:

(1) Z = depth of compartments

(2) D = lengths of flow segments between adjacent compartments in each column profile

(3) ZT = total depth of profile to impervious bottom boundary

(4) IVWA,B,C,D,E, = initial volume of water in each compartment and in each of the columns designated A, B, C, D, E, starting from the upslope side. A gradually increasing water content with depth is assumed initially.

Note: In the program given, all columns are equal in terms of width, initial moisture distribution, soil moisture characteristic, and hydraulic properties, slope, water-table depth, surface properties, etc. In principle, however, different properties can be assigned to each column as necessary to include lateral heterogeneity. Vertical heterogeneity can be included in each column as shown in Chapter 3.

DYNAMIC Section

Rainfall rate (RAIN) is determined by interpolation of the RAINTB function. Water volume in each compartment in each column (VWA, VWB, VWC, VWD, VWE) is determined by integration of the net flux for each compartment (designated NQA1,19 for the first column and B, C, D, E for the next four), with the initial volumes given by the IVW values computed in the INITIAL section.

Depth of surface water (HO) is obtained by integration of the time-change of surface water depth (DHDT). Cumulative runoff CO is obtained by integration of the runoff rate (RO) occurring across each columnar boundary. The volumetric wetness W (doubly subscripted for each compartment in each column) is the ratio of the respective water volume VW to the respective compartment thickness L.

The DO 120 loop computes for each column, and the nested DO 110 loop computes for each compartment in each column, the values of pressure or matric potential head (P) of soil water, the hydraulic head (H), the hydraulic conductivity (K),

Figure 4.4. CSMP listing for computing the water dynamics of a uniform sloping field, including infiltration and runoff, as well as flow in the unsaturated zone and below the water table.

```
TITLE              RAIN - INFILTRATION - OVERLAND FLOW

*          UNITS
*     KG = KILOGRAMS
*     M = METERS
*     S = SECONDS

*                          GLOSSARY OF SYMBOLS

* ALF     = COMPOSITE CONSTANT OF MANNINGS EQUATION
* AK      = AVERAGE HYDRAULIC CONDUCTIVITY BETWEEN LAYERS (M/S)
* CF      = CUMULATIVE INFILTRATION (M)
* CO      = CUMULATIVE RUNOFF (M**2)
* CQL     = CUMULATIVE LATERAL FLOW OF GROUNDWATER (M**2)
* CRAIN   = CUMULATIVE RAINFALL (M)
* D       = DISTANCE OF FLOW BETWEEN ADJACENT LAYERS (M)
* DHDT    = TIME-CHANGE OF SURFACE-WATER DEPTH (M/S)
* EVAP    = EVAPORATION RATE (M/S)
* F       = INFILTRATION RATE (M/S)
* G       = INDEX OF TOPMOST LAYER UNDER WATER TABLE
* H       = HYDRAULIC POTENTIAL HEAD (M)
* HO      = HEIGHT OF SURFACE WATER (M)
* I       = INDEX OF DEPTH
* IVW     = INITIAL VOLUME OF WATER PER LAYER (M)
* IW      = INITIAL VOLUMETRIC WETNESS (DIMENSIONLESS)
* J       = INDEX OF COLUMN
* K       = HYDRAULIC CONDUCTIVITY (M/S)
* L       = THICKNESS OF LAYERS (M)
* M       = EXPONENTIAL CONSTANT OF MANNINGS EQUATION
* N       = MANNINGS CONSTANT (SURFACE CHARACTERISTIC PARAMETER)
* NC      = NUMBER OF COLUMNS
* NF      = INFILTRABILITY (M/S)
* NL      = NUMBER OF LAYERS
* NQA     = NET FLUX IN PROFILE A (M/S). SAME FOR PROFILES B,C,D,E
* P       = PRESSURE POTENTIAL HEAD (M)
* PET     = POTENTIAL EVAPORATION RATE (M/S)
* Q       = FLUX OF WATER IN PROFILE (M/S)
* QL      = LATERAL FLOW RATE OF GROUNDWATER (M**2/S)
* RAIN    = RAINFALL RATE (M/S)
* RO      = RUNOFF RATE (M**2/S)
* S       = SLOPE OF SOIL SURFACE (DIMENSIONLESS)
* SK      = SATURATED HYDRAULIC CONDUCTIVITY (M/S)
* VWA     = VOLUME OF WATER IN EACH LAYER OF PROFILE A (M) (SAME
*           FOR PROFILES, B,C,D,E)
* W       = VOLUMETRIC WETNESS (DIMENSIONLESS)
* WT      = WATER TABLE HEIGHT ABOVE IMPERVIOUS PLANE (M)
* WTT     = WATER TABLE HEIGHT AT DRAIN (M)
* X       = WIDTH OF EACH COLUMN (M)
* Z       = DEPTH OF MIDPOINT OF EACH LAYER (M)
* ZT      = TOTAL DEPTH OF PROFILE TO IMPERVIOUS PLANE (M)
```

```
STORAGE      L(20),Z(20),H(20),K(20),AK(20),NF(20),D(20),WT(9),G(9),QL(9)
/      DIMENSION VWA(20),VWB(20),VWC(20),VWD(20),VWE(20),W(20,20),CO(20)
/      DIMENSION NQA(20),NQB(20),NQC(20),NQD(20),NQE(20),HO(20),DHDT(20)
/      DIMENSION RO(20),CF(20),IVWA(20),F(20),Q(20,20),IVWB(20),IVWC(20)
/      DIMENSION IVWD(20),IVWE(20),P(20,20),EVAP(20)
/      EQUIVALENCE (VWA1,VWA(1)),(VWB1,VWB(1)),(VWC1,VWC(1)),(RO1,RO(1))
/      EQUIVALENCE (VWD1,VWD(1)),(VWE1,VWE(1)),(IVWA1,IVWA(1)),(CO1,CO(1))
/      EQUIVALENCE (IVWB1,IVWB(1)),(IVWC1,IVWC(1)),(IVWD1,IVWD(1))
/      EQUIVALENCE (IVWE1,IVWE(1)),(NQA1,NQA(1)),(NQB1,NQB(1)),(F1,F(1))
/      EQUIVALENCE (NQC1,NQC(1)),(NQD1,NQD(1)),(NQE1,NQE(1)),(CF1,CF(1))
/      EQUIVALENCE (HO1,HO(1)),(DHDT1,DHDT(1))
FIXED I,NL,NLL,NC,NCC,G ,J,JJ

INITIAL

NOSORT
PARAM NL=19,NC=5,X=50.,IW=.2,SK=.7E-5,MINPOT=-1000.,M=2.,N=.03,A=1.
PARAM S=.02, WTT=.16
             NLL=NL+1
             NCC=NC+1
             ALF=A*S**.5/N
TABLE L(1-19)=.02,.04,16*.06,.10
             Z(1)=.5*L(1)
      DO  10 I=1,NL
             IVWA(I)=(IW+AMIN1(.02*I,.285))*L(I)
             IVWB(I)=(IW+AMIN1(.02*I,.285))*L(I)
             IVWC(I)=(IW+AMIN1(.02*I,.285))*L(I)
             IVWD(I)=(IW+AMIN1(.02*I,.285))*L(I)
          10 IVWE(I)=(IW+AMIN1(.02*I,.285))*L(I)
             D(1)=Z(1)
      DO  12 I=2,NL
             D(I)=.5*(L(I-1)+L(I))
          12 Z(I)=Z(I-1)+D(I)
             ZT=Z(NL)+.5*L(NL)
FUNCTION SUCTB=................................
FUNCTION CONDTB=...............................
FUNCTION RAINTB=(0.,0.),(3600.,1.E-5),(10800.,1.E-5),(14400.,0.)
             PET=.01/86400.

DYNAMIC

NOSORT
             RAIN=AFGEN(RAINTB,TIME)
             VWA1=INTGRL(IVWA1,NQA1,19)
             VWB1=INTGRL(IVWB1,NQB1,19)
             VWC1=INTGRL(IVWC1,NQC1,19)
             VWD1=INTGRL(IVWD1,NQD1,19)
             VWE1=INTGRL(IVWE1,NQE1,19)
             HO1=INTGRL(.00001,DHDT1,5)
             CO1=INTGRL(0.,RO1,6)
      DO 100 I=1,NL
             W(1,I)=VWA(I)/L(I)
             W(2,I)=VWB(I)/L(I)
             W(3,I)=VWC(I)/L(I)
             W(4,I)=VWD(I)/L(I)
         100 W(5,I)=VWE(I)/L(I)
      DO 120 J=1,NC
      DO 110 I=1,NL
             P(J,I)=-AFGEN(SUCTB,W(J,I))
             K(I)=AFGEN(CONDTB,W(J,I))
             H(I)=P(J,I)-Z(I)
             IF (I.EQ.1) GO TO 110
             AK(I)=(K(I-1)*L(I-1)+K(I)*L(I))/(L(I-1)+L(I))
             Q(J,I)=(H(I-1)-H(I))*AK(I)/D(I)
             Q(J,NLL)=0.
         110 CONTINUE
```

```
              NF(J)=(0.-H(1))*.5*(SK+K(1))/D(1)
              G(J)=NL
       DO 115 I=1,NL
              JJ=NL-I+1
              IF (P(J,JJ).GE.0.) G(J)=JJ-1
          115 CONTINUE
          120 CONTINUE
       DO 125 J=1,NC
          125 IF (RAIN.GT.0.) EVAP(J)=0.
              IF (RAIN.GT.0.) GO TO 145
       DO 140 J=1,NC
              EVAP(J)=PET
              IF (P(J,1).LE.MINPOT) EVAP(J)=AMIN1(PET,-Q(J,2))
          130 Q(J,1)=-EVAP(J)
          140 CONTINUE
          145 CONTINUE
       DO 200 J=1,NC
              WT(J)=ZT-Z(G(J)+1)+.5*L(G(J)+1)
              F(J)=NF(J)
              IF (HO(J).LE.0.) F(J)=AMIN1(NF(J),RAIN)
              Q(J,1)=F(J)
          200 CONTINUE
       DO 210 I=1,NL
              NQA(I)=Q(1,I)-Q(1,I+1)
              NQB(I)=Q(2,I)-Q(2,I+1)
              NQC(I)=Q(3,I)-Q(3,I+1)
              NQD(I)=Q(4,I)-Q(4,I+1)
              NQE(I)=Q(5,I)-Q(5,I+1)
          210 CONTINUE
              QL(1)=0.
       DO 220 J=2,NC
          220 QL(J)=(((WT(J-1)-WT(J))/X)+S)*SK*.5*(WT(J-1)+WT(J))
              QL(NCC)=((WT(NC)-WTT)/(.5*X)+S)*SK*(WT(NC)-WTT)
              NQA(G(1))=NQA(G(1))+(QL(1)-QL(2))/X
              NQB(G(2))=NQB(G(2))+(QL(2)-QL(3))/X
              NQC(G(3))=NQC(G(3))+(QL(3)-QL(4))/X
              NQD(G(4))=NQD(G(4))+(QL(4)-QL(5))/X
              NQE(G(5))=NQE(G(5))+(QL(5)-QL(6))/X
              TJP1=(AMAX1(HO(1)+HO(2),0.))**M
              DHDT(1)=(-ALF*TJP1/(X*2.**M))+RAIN-F(1)
       DO 230 J=2,NC
              TJM1=(AMAX1(HO(J)+HO(J-1),0.))**M
              TJP1=(AMAX1(HO(J)+HO(J+1),0.))**M
          230 DHDT(J)=(ALF/(X*2.**M))*(TJM1-TJP1)+RAIN-F(J)
              HO(NCC)=HO(NC)+.5*(HO(NC)-HO(NC-1))
              RO(1)=0.
       DO 250 J=2,NC
              THO=AMAX1((HO(J-1)+HO(J)),0.)
          250 RO(J)=ALF*(THO*.5)**M
              THO=AMAX1(HO(NCC),0.)
              RO(NCC)=ALF*THO**M
              CRAIN=INTGRL(0.,RAIN)
              CF1=INTGRL(0.,F1,5)
              RO6=RO(6)
              CO6=INTGRL(0.,RO6)
              QL6=QL(6)
              CQL6=INTGRL(0.,QL6)

TERMINAL

TIMER FINTIM=86400.,OUTDEL=900.,DELMIN1.E-6
PRINT (optional)
PRTPLT (optional)
METHOD RKS
END
STOP
```

the average conductivity for flow between adjacent layers (AK), and the flow rate between layers (Q) from Darcy's law with the bottom flow (Q(J,NLL)) being zero because of the impervious boundary there. The infiltrability of each column (NF(J)) is computed under the assumption that the pressure of water at the soil surface is zero (*i.e.*, atmospheric pressure). Note that index J refers to the ordinal number of each column and index I refers to ordinal number of each compartment within each column.

The DO 115 loop, also nested inside the DO 120 loop, is a search procedure to identify the topmost compartment (re-indexed G) which is at a pressure potential greater than or equal to zero. The vertical location of the top of that compartment is then taken to be the water-table height in each column (WT(J)).

The DO 125 and 125 statements set the evaporation rate at zero during a rainstorm. While there is no rain (DO 140), evaporation takes place at the potential rate (PET) until the matric potential (P) in the surface compartment of each column falls to the air-dry value (MINPOT), after which evaporation may fall below the potential rate and equals the supply rate from the profile below (-Q(J,2)).

During a rainstorm (statement 145) infiltration rate is equal to infiltrability unless there is no free water at the soil surface, in which case infiltration rate is equal to either infiltrability or to rainfall rate, whichever is the lesser (as determined by the AMIN1 condition). The DO 210 loop then computes the net fluxes for each compartment in each column.

Next, the lateral flows below the water table are calculated. The upper boundary (taken to be the crest of the water table) is set at zero: QL(1) = 0. DO 220 calculates groundwater flows, QL(J), between laterally adjacent columns on the basis of Dupuit-Forchheimer assumptions. Lateral flow through the downhill boundary QL(NCC), is computed similarly, except that the water table at the outlet, WTT, is taken to be constant. The net flow of the topmost compartment of the saturated zone in each column, *i.e.*, the compartment just below the water table [NQA(G(1)), NQB(G(2)), etc.] is then corrected to account for lateral groundwater flow. Thus, if a particular column has greater outflow than inflow of groundwater during any particular period, the net outflow is removed from the topmost saturated compartment, and, unless there is sufficient replenishment from the profile above, that compartment may become unsaturated and cause the water table to fall from its top boundary to its bottom boundary.

The next two statements compute the rate of change of surface-water depth for the first column, using the kinematic wave equation, assuming that there is no surface-water inflow into this uphill column [RO(1) = 0]. The DO 230 loop then computes the corresponding rates of change of surface water depth for the other columns. The water depth over the last boundary (downslope edge of the field) is determined by extrapolation in the HO(NCC) statement. The runoff rates, RO(J), are computed by Manning's equation in the DO 250 loop. Cumulative rain (CRAIN) and cumulative infiltration (CFl) are obtained by integration of the rainfall and infiltration rates over time. The cumulative lateral flows (CQL) and cumulative evaporation values (EVAP) are also obtained by integration with respect to time of the appropriate rates.

The TERMINAL section should by now be self-explanatory.

D. Results of Simulation Trials

To test and demonstrate the capabilities of the model, a number of simulation trials were conducted. These will be described in turn.

Rainfall Over an Impervious Slope

A heavy rainstorm of 6 hour duration (2 hours of increasing intensity, 2 hours of steady intensity at 72 mm/hour, and 2 hours of diminishing intensity), totaling 288 mm, was simulated.

Figure 4.5 shows the appearance and mean depth of surface water excess over the first 50 m segment of the slope as a function of time for four gradients: 1%, 2%, 10%, and 20%. In each case, the thickness of the layer of flowing water increases with rain intensity, levels out as the rain becomes steady, and decreases as rain intensity diminishes. However, the recession of surface water lags somewhat behind the cessation of the rain, as is evident in the right-side "tails" exhibited by the curves of Figure 4.5.

Comparing the steady-phase values of surface-water thickness, we note that increasing slope steepness results in decreasing mean thickness of the running water. This is obviously due to the increased velocity of overland flow. However, the decrease of surface-water thickness is much less than proportional to slope steepness. Thus, a twenty-fold increase in the slope (from 1% to 20%) causes only a two-fold decrease in surface-water thickness. This is, of course, a consequence of the use of the Manning's equation, as described in section B of this chapter.

The increase of surface-water thickness with distance downslope is shown in Figure 4.6. At the lower edge of our field, 225 m downslope from the crest, the steady-state thickness of the flowing sheet of water is about 30 mm in the nearly level (1% slope) impervious field but only about 14 mm in the

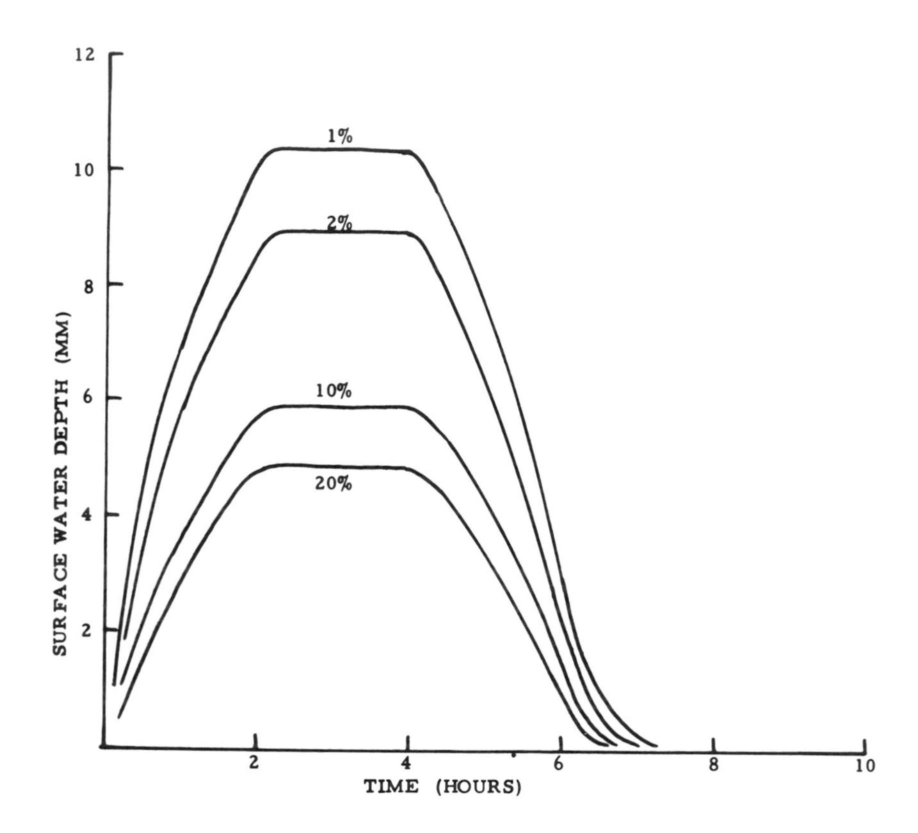

Figure 4.5. Computed values of mean surface water excess over an impervious surface during a rainstorm of 288 m for four slopes: 1%, 2%, 10%, 20%.

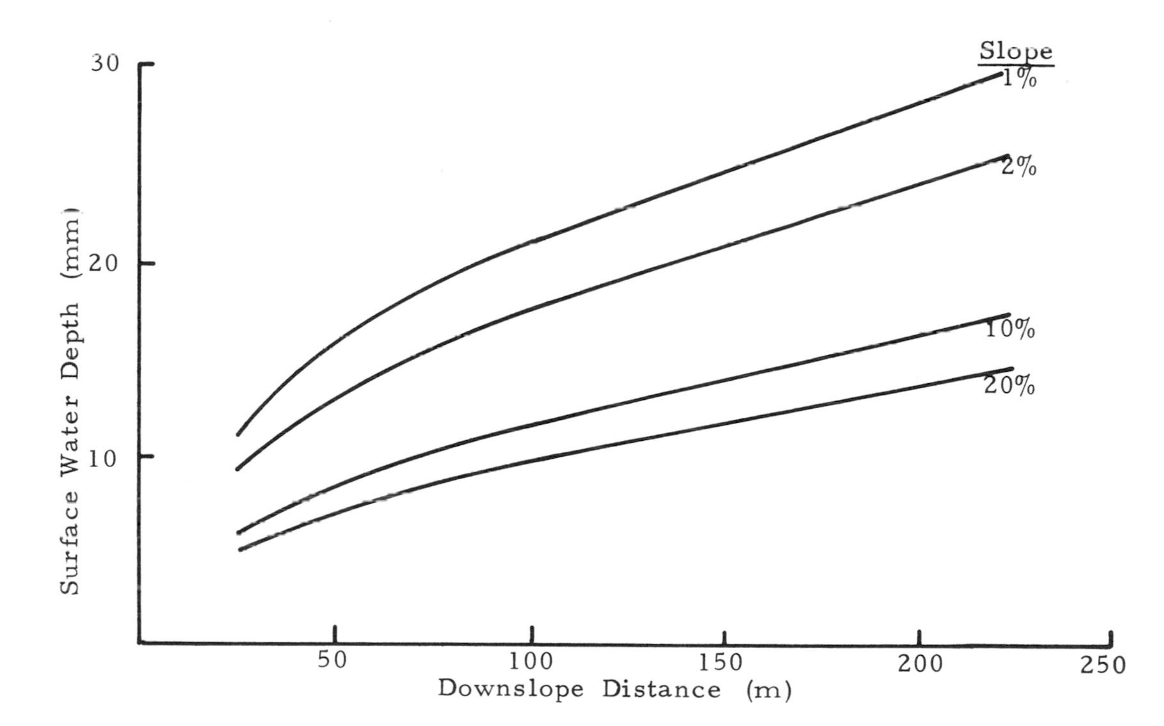

Figure 4.6. Surface-water thickness over a sloping impervious surface as function of downslope distance during the steady phase of a rainstorm with an intensity of 72 mm/hour.

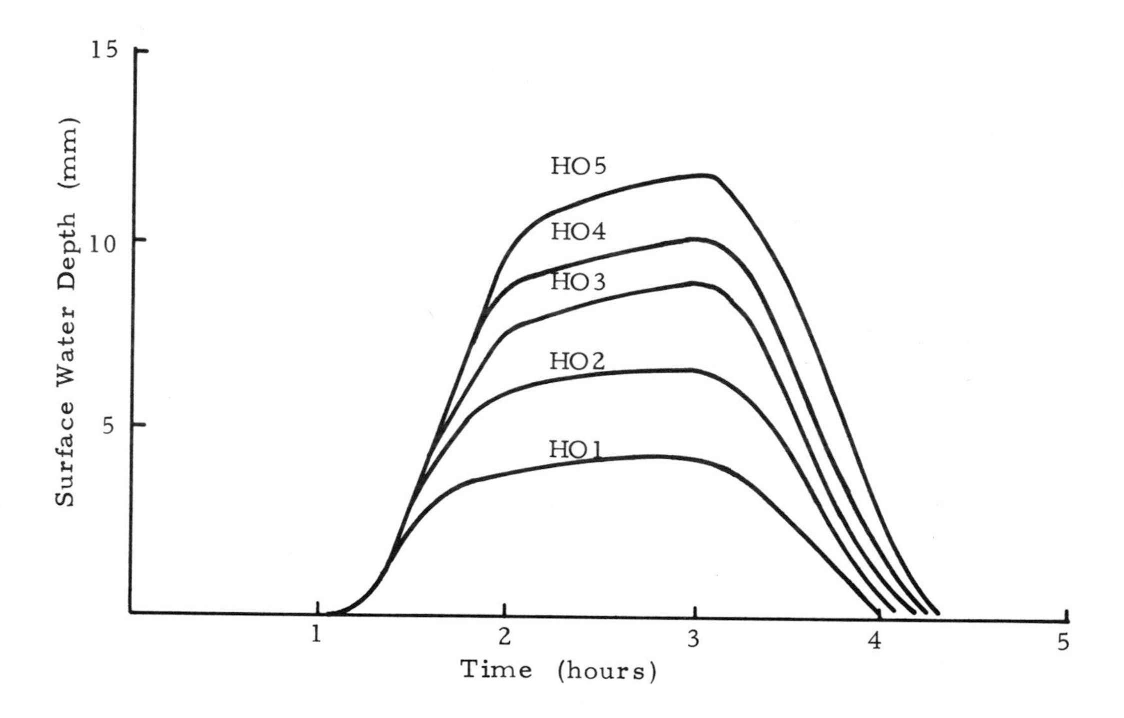

Figure 4.7. Development of surface water excess over a field of uniform clay at a slope of 4% during a 108-mm rainstorm lasting 4 hours, with a two-hour intensity of 36 mm/hour.

steepest (20% slope) field. In all cases, surface water increases in thickness somewhat less than proportionately with downslope distance, with approximately a three-fold increase over the range of distance from the first segment of the slope to the last.

A trial was also conducted to determine the sensitivity of the model to variation of slope-segment width. For an impervious surface, the thickness of flowing surface-water can be determined directly from the continuous theory presented in section B of this chapter. The results of such calculations in comparison with the results obtained from our segmented-slope numerical model, are shown in Table 4.1.

Table 4.1. Surface-water thickness (mm) at a distance of 25 m downslope from the crest of an impervious sloping field under a steady rainfall of 72 mm/hour.

Slope	Theoretical	Slope-Segment Width	
		10m	50m
1%	10.1	10.3	11.2
2%	8.4	8.7	9.4
10%	5.6	5.8	6.3
20%	4.7	4.9	5.3

It is seen that the estimate of surface-water thickness worsens with increasing width of the assumed slope segment. However, the errors associated with 50 m wide segments are considered tolerable at the present stage of the model's development. Smaller and hence more numerous columns can of course be programmed but require more computer time for the simulation.

Rainfall Over a Field of Uniform Clay

The development of surface water excess over an infiltrating field of uniform clay at a slope of 4 percent is shown in Figure 4.7. Here the rainstorm totaled 108 mm, with 1 hour of increasing intensity, 2 hours of steady intensity at 36 mm/hour, and 1 hour of decreasing intensity. Surface water is seen to increase in thickness during the increasing intensity and the steady intensity phases of the simulated storm. Evidently, soil infiltrability was in the decreasing phase throughout the duration of the storm, which was not long enough to bring about the steady infiltration phase.

The same pattern is reflected in the rainfall and runoff hydrograph shown in Figure 4.8. Here the difference between the rainfall and runoff rates indicates the pattern of infiltration which at first equals the rainfall rate but begins to fall progressively below it after 1½ hours. The total amount of runoff, as determined by integration, was about 35 percent of the total rainfall over the field.

A similar study of rain-infiltration in a simulated field of uniform loam produced no runoff at all. The hydraulic properties of both soils were the same as those elucidated in Chapter 3 of the monograph.

Rainfall Over a Composite Field of Clay and Loam

The next set of simulation trials involved laterally heterogeneous fields of loam and clay. Two hypothetical fields were compared: the one with an upslope column of loam and four downslope columns of clay, and the other with an upslope column of clay and four downslope columns of loam.

The rate of runoff discharge from one column to the next as function of downslope distance is shown in Figure 4.9 for the various fields being compared. These fields are designated as follows: "C" for the uniform clay, "L" for uniform loam (curve not shown, since there was no runoff), "C/L" for uphill clay and downhill loam, and "L/C" for uphill loam and downhill clay. The curves show that runoff discharge increased linearly with downslope distance in the case of the clay and loam-over-clay fields, the latter simply producing less runoff owing to the failure of the uppermost loam column to contribute any surface water. The curve for the clay-over-loam field indicates that the runoff generated by the clay segment of the slope is partially re-absorbed by the loam during overland flow so that the overall discharge diminishes with downslope distance. If the loamy segment of the slope were longer, there might be no runoff at all.

The total quantities of runoff and of lateral groundwater drainage discharged from the variously constituted fields are shown in Table 4.2. The fields which were entirely or predominantly clay produced more runoff but considerably less drainage, owing to the smaller hydraulic conductivity of the saturated clay as compared to the saturated loam (namely, 2×10^{-6} m/sec versus 7×10^{-6} m/sec, respectively). The quantities of groundwater flow in our simulated clayey fields were some three orders of magnitude smaller than the quantities of runoff, but were of course greater than runoff in the loam field. Over a longer period of time marked by extended periods of no rainfall, the quantities of drainage might become relatively significant even in clayey fields, while in sandy fields they are likely to

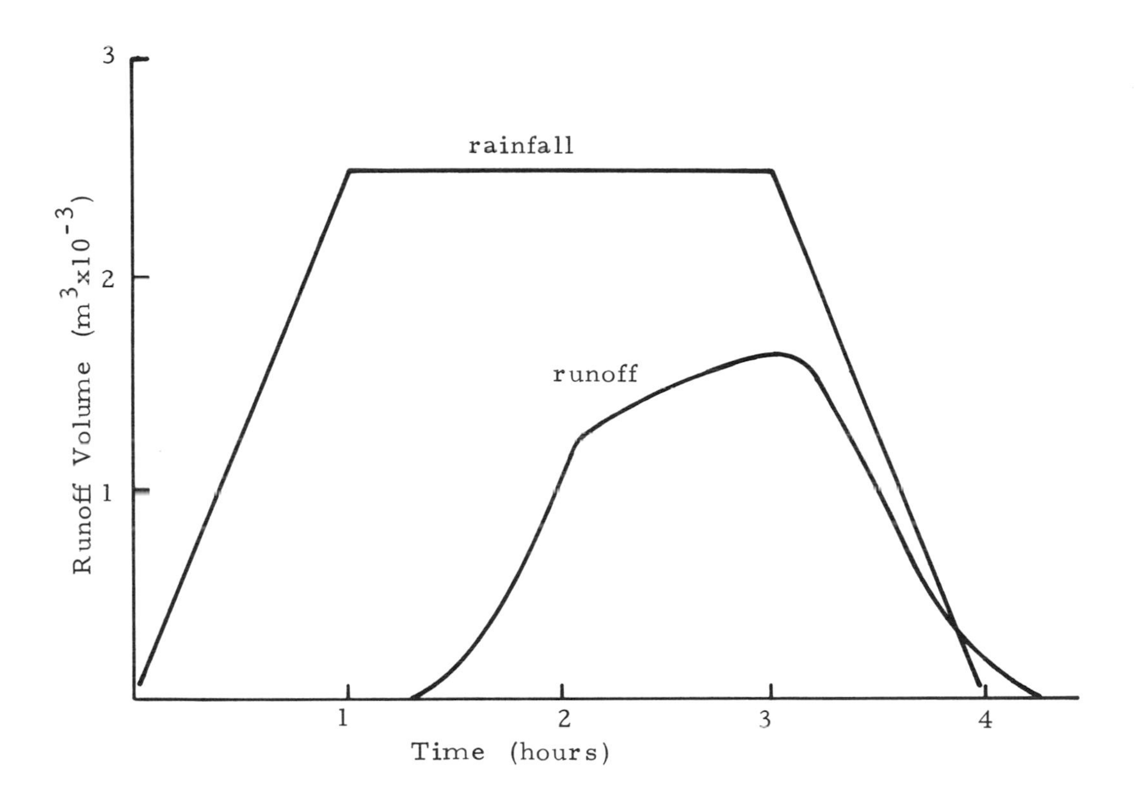

Figure 4.8. Rainfall and runoff hydrograph for a sloping field of clay under a rainstorm of 108 mm. Note that the rainfall and runoff are given in terms of water volume per time. 1 mm over a 1 m wide, 250 m stretch of slope is equivalent to 0.25 m^3.

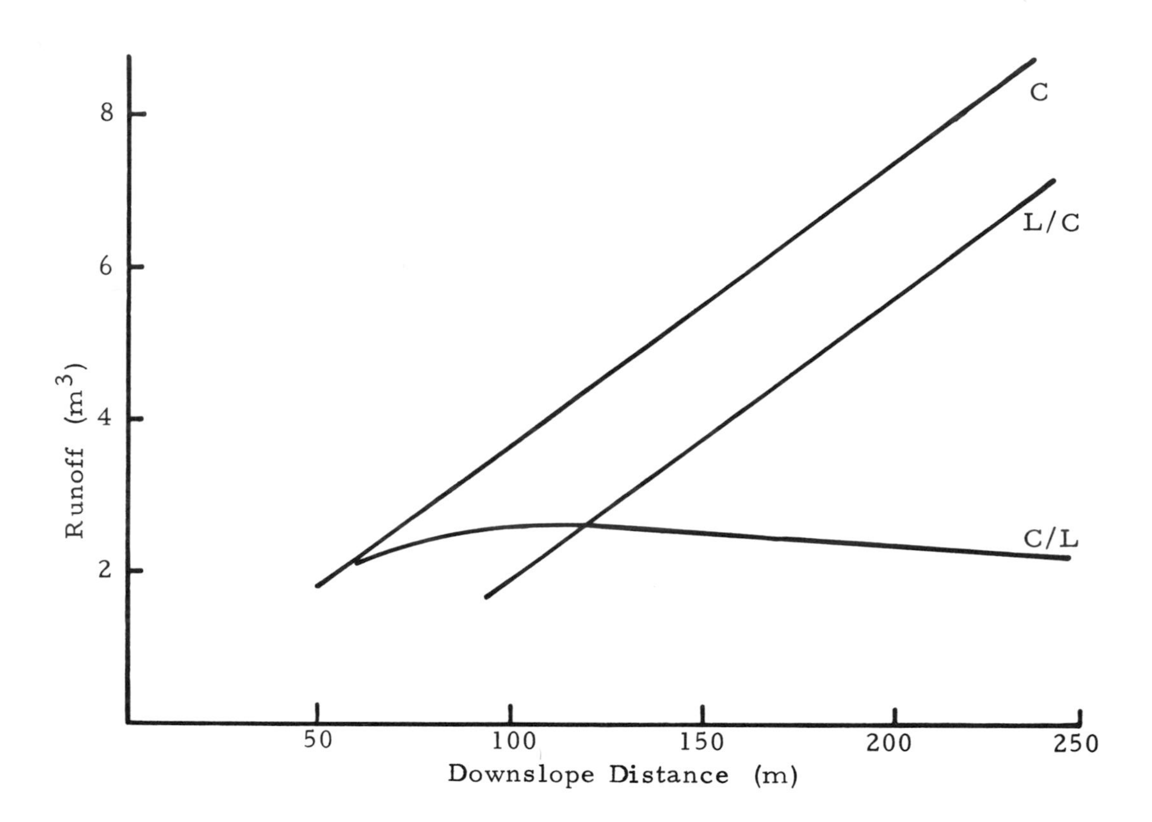

Figure 4.9. Runoff discharge as function of downslope distance for fields of various composition during the steady phase of a 108-mm rainstorm with 2-hours at an intensity of 36 mm/hour. C, L, refer to clay and loam, respectively.

predominate over runoff at all times.

Table 4.2. Overall quantities of runoff and lateral groundwater drainage discharged during one day by variously constituted sloping fields measuring 250 m in downslope length and 1 m in width. Data in cubic metres.

Field	Total Runoff	Total Drainage
Uniform clay	9.66	1.47×10^{-3}
Uniform loam	—	8.77×10^{-3}
Clay over loam	2.26	7.89×10^{-3}
Loam over clay	7.88	1.49×10^{-3}

To better interpret the data of Table 4.2, note that the simulation extended over a period of one day only, during which a sizable rainstorm had been made to occur. The total amount of rainfall (108 mm over 250 m^2 of the field) amounted to 27 cubic metres.

Discussion

The simulation trials described above, though preliminary, illustrate the capability of the model to describe the overall hydrological balance of a sloping field which can be represented by a two-dimensional cross section. As mentioned earlier, the model can easily be adapted to represent fields of variable slope and textural composition. Plant activity and root extraction of soil water can be included. A solute transport model can also be grafted on the hydraulic flow model to describe the fate of salts and their possible migration between the soil and the groundwater. Vertical heterogeneity can be taken into account by the method described in our preceding chapter. The possibility of including lateral heterogeneity effects is a particularly important feature of our model. Recently shown to be an inescapable characteristic of agricultural soils, its possible effect in principle on the overall water regime of a field had not yet been analyzed.

However, as against its promising features, our model has a number of serious shortcomings. Its disregard of lateral flow in the so-called unsaturated zone (above the water table) can cause errors where intermediate clay layers occur in the soil and may cause the formation of temporary, perched water tables , particularly where such layers are slanted rather than horizontal. Moreover, the routing of runoff using Manning's equation disregards non-sheet flow, such as concentrated flow in rills or microgulleys, as well as raindrop impact effects. A two-dimensional model also fails to account for the generally three-dimensional convergent flow pattern typical of watersheds draining into streams. Finally, the complexity of the model, as programmed in CSMP, makes it cumbersome to run and may consume more time on the computer than is generally feasible. More sophisticated and efficient programming techniques (*e.g.*, the use of the finite-element approach) are necessary for a model which is to incorporate vertical and horizontal soil heterogeneities, etc. For the moment, however, suffice it to say that the model described, being an extension of the traditional one-dimensional models of soil physics, has proven itself to be internally consistent and logically sound, and ought certainly to be developed further and to be tested in the field.

V. MOISTURE EXTRACTION BY ROOT SYSTEMS AND THE CONCURRENT MOVEMENT OF WATER AND SALT IN THE SOIL PROFILE

A. Description of the Problem

Plants growing in the field, particularly in arid regions, are required by the prevailing climatic environment to transpire large quantities of water. To grow successfully, each plant must achieve a water economy such that the demand made upon it is balanced by the rate at which it can extract water from the soil. To optimize the management and utilization of soil moisture, we need detailed knowledge and fundamental understanding of the processes involved.

Plants live in two realms, the atmosphere and the soil, in each of which the physical conditions vary continuously. The exact manner in which various plants respond to the combination of conditions prevailing in the atmosphere and in the soil yet remains to be elucidated quantitatively. Pioneering advances in this direction have been made by Philip (1957), Gardner (1960), Cowan (1965), and others.

A major problem encountered in any attempt at an exact physical description of soil-water uptake by plant roots is the inherently complicated space-time relationship involved. Roots grow in different directions and spacings, and at different rates. Also, they exhibit sectional differences in absorptive activity, depending upon age and location. Moreover, old roots die while new ones proliferate at a rate which depends on the physical and chemical environment (*e.g.*, temperature, moisture, nutrients, salinity, aeration, etc.) as well as on physiological factors. How the root system of a plant senses the root zone as a whole and integrates its response so as to utilize soil moisture to best advantage has long been a subject of great interest. One classical view (Wadleigh 1946) was that the root system adjusts its water withdrawal pattern so as to maintain the total soil moisture potential constant throughout the root zone. On the other hand, an often-observed pattern of water withdrawal is such that the top layer is depleted first and the zone of maximal extraction moves gradually into the deeper layers.

Since the soil usually extends in depth considerably below the zone of root activity, it is of interest to establish how the pattern of soil water extraction by roots relates to the pattern of water flow within, through, and below the root zone. Some drainage through the root zone is considered necessary to prevent deleterious accumulation of salts, particularly in the arid zone; yet excessive drainage might involve unnecessary loss of nutrients as well as of water. If a ground water table is present at a shallow depth, it can contribute to the supply of water to the root zone by upward capillary flow, but it might also infuse the root zone with harmful salts. Considerable upward capillary flow is possible even in the absence of a water table, if the depleting root zone is underlain by moist layers with sufficient storage and conductivity. In fact, the opposite processes of downward flow and capillary rise can occur in an alternating pattern at varying rates so that the net outflows or inflows of water and of salts for the root zone as a whole can only be determined by integrating the fluxes taking place through the bottom of the root zone continuously over an extended period such as a growing season.

The current approach to plant water uptake is based on recognition that the field environment forms a unified system which Philip (1966) has called the "SPAC" (for "soil-plant-atmosphere continuum"). In this system, water flows in a "transpiration stream" down a gradient of potential energy from soil to root to stem to leaf, whence it evaporates and diffuses out to the atmosphere. Employing the analogy of Ohm's law for an electric current through a series of resistors, van den Honert (1948) represented the transpiration stream as a catenary process through successive segments, in each of which the flux (q) is proportional to a potential difference ($\Delta\Phi$) and inversely proportional to a resistance (R):

$$q = - \frac{\Delta\Phi}{R} \tag{5.1}$$

In the case of the soil segment of the transpiration stream, the resistance varies greatly in time and space, as it depends on soil and root-system hydraulics. To model soil water uptake in quantitative physical terms, two alternative approaches have been tried: (1) The *microscopic-scale approach* (*e.g.* Gardner 1960; Molz *et al.* 1968; Lambert and Penning de Vries 1973; Hillel *et al.* 1975b), which analyzes the radial flow of water to individual roots, considered to be line or narrow-tube sinks regularly spaced in the soil; and (2) the *macroscopic-scale approach* (*e.g.*, Whisler *et al.* 1968; Molz and Remson 1970, 1971; Nimah and Hanks 1973), which regards the root system in its entirety as a diffuse sink permeating the soil continuously, though not necessarily at uniform strength throughout the root zone. The relative merits of the two approaches were discussed elsewhere (Hillel *et al.* (1975b).

Some of the previously published macroscopic-scale models of soil water extraction by root systems assumed steady-state flow from a water table located at some constant depth, or negligible root resistance (*i.e.*, all roots having uniform water potential) or disregarded osmotic effects. Others imposed arbitrarily restrictive or empirically fitted equations for the soil's hydraulic conductivity function, for the extraction rate, or for the effective root resistance value; or were in some other way nonmechanistic. The model described herein is an attempt to formulate the process, insofar as seems possible at present, in terms of basic physical mechanisms expressible as transport equations for water and solutes in and through the soil profile and the root system. The model can provide for the possibility that the soil profile itself, as well as the root system, is nonuniform in depth. Furthermore, the model attempts to calculate rates and overall quantities for the drainage of water and leaching of solutes beyond the root zone.

It is our hope that an integrated approach of the sort illustrated herein will contribute to a more complete understanding of soil and climatic factors as they might combine to affect plant water status and uptake, as well as water and solute transmission through the soil. Greater understanding in quantitative terms of the dynamics of soil moisture availability to plants is indeed necessary in the continuing effort to improve and optimize the agronomic, hydrologic, and environmental aspects of soil-water management.

B. Governing Equations

The vertical transient-state flow of water in a stable and uniform segment of the root zone can be described by the following equation:

$$\frac{\partial \theta}{\partial t} = - \frac{\partial}{\partial z} [K(\theta) \frac{\partial (\Psi + z)}{\partial z}] - S_W \qquad (5.2)$$

in which θ is volume wetness, t time, z depth, $K(\theta)$ hydraulic conductivity (a function of wetness), Ψ matric suction head, and S_W is a sink term representing extraction by plant roots.

The rate and direction of solute movement in a soil system depends largely on the pattern of water movement, but is also affected by diffusion and hydrodynamic dispersion (Nielsen and Biggar 1962).

If the latter effects are negligible, solute flow by convection can be formulated as:

$$J_C = qc = \bar{V}\theta c$$

where J_C is the flux density of solute, q the flux density of water, c the concentration of solute in the flowing water and $\bar{V}$ the average velocity of flow.

The rate of diffusion of a solute (J_d) in bulk water at rest is related by Fick's law to the concentration gradient:

$$J_d = D_o(\partial c/\partial x) \tag{5.4}$$

in which D_o is the diffusion coefficient.

In the soil the effective diffusion coefficient D_s is decreased owing to the fact that the liquid phase occupies only a fraction of soil volume, and also owing to the tortuous geometry of the path:

$$D_s = D_o\theta\xi \tag{5.5}$$

in which ξ, the tortuosity, is an empirical factor smaller than unity, which can be expected to decrease with decreasing θ.

In addition to molecular diffusion, convective flow generally causes hydrodynamic dispersion, an effect which results from the microscopic nonuniformity of flow velocity in the various pores. Thus, a sharp boundary between two miscible solutions becomes increasingly diffuse about the mean position of the front. The magnitude of the dispersion coefficient, D_h, has been found to depend linearly on the average flow velocity, $\bar{V}$ (Bresler 1973):

$$D_h = \alpha\bar{V} \tag{5.6}$$

where α is an empirical coefficient.

The diffusion and dispersion effects can be combined with the convective transport equation to give the overall flux of solute, J:

$$J = -(D_h + D_s)(\partial c/\partial x) + \bar{V}\theta c \tag{5.7}$$

With continuity brought into consideration, one-dimensional transient movement of a non-interacting solute in soil becomes:

$$\frac{\partial(\theta c)}{\partial t} = \frac{\partial}{\partial z}\left(D_a \frac{\partial c}{\partial z}\right) - \frac{\partial(qc)}{\partial z} - S_s \tag{5.8}$$

where c is concentration of the solute in the soil solution, q is convective flux of the solution, D_a is a combined diffusion and dispersion coefficient, and S_s is a sink term for the solute representing root absorption, precipitation, volatilization, or any other mechanisms by which the solute may be removed from the flowing solution.

The rate of extraction of water from a unit volume of soil can be represented in the following way:

$$S_w = \frac{\phi_{soil} - \phi_{plant}}{R_{soil} + R_{roots}} \quad (5.9)$$

Herein ϕ_{soil} is the total potential of soil water, being the sum of the matric (ϕ_m), gravitational ($\phi_g = z$) and osmotic (ϕ_o) potentials, all of which are expressible in head units:

$$\phi_{soil} = \phi_m + \phi_g + \phi_o \quad (5.10)$$

The hydraulic resistance to flow in the soil toward the roots was expressed by Gardner (1964) as inversely proportional to the hydraulic conductivity (K) (a function of the soil's wetness or matric potential) and to the total length of active roots (L) in the unit volume of soil:

$$R_{soil} = 1/BKL \quad (5.11)$$

Herein, B is an empirical constant, which can be taken to represent a root-length activity factor.

The term ϕ_{plant} is the plant water potential at a point, presumably at the base of the stem, where all roots converge and the plant emerges from the soil with a single water potential which we choose to call the "crown potential" (hereafter to be designated ϕ_c).

The hydraulic resistance of the roots (R_{roots}) can be taken to be the sum of a resistance to absorption and a resistance to conduction, the latter being a function of the depth of any particular group of roots. Problems associated with the characterization of root resistance will be described more fully in the Discussion section of this chapter.

The flow rate (q_r), delivered by the roots from any particular layer i in the soil to the crown can be taken as the ratio of the difference in potential between that soil layer and the crown to the total hydraulic resistance encountered:

$$(q_r)_i = \frac{(\phi_s)_i - \phi_c}{(R_r)_i + (R_s)_i} \quad (5.12)$$

where $(\phi_s)_i$ is the soil moisture potential, $(R_r)_i$ the resistance of the roots, and $(R_s)_i$ is the hydraulic resistance of the soil.

The total extraction rate (Q) from all volume elements or layers of soil, equal to the transpiration rate, is a sum of the

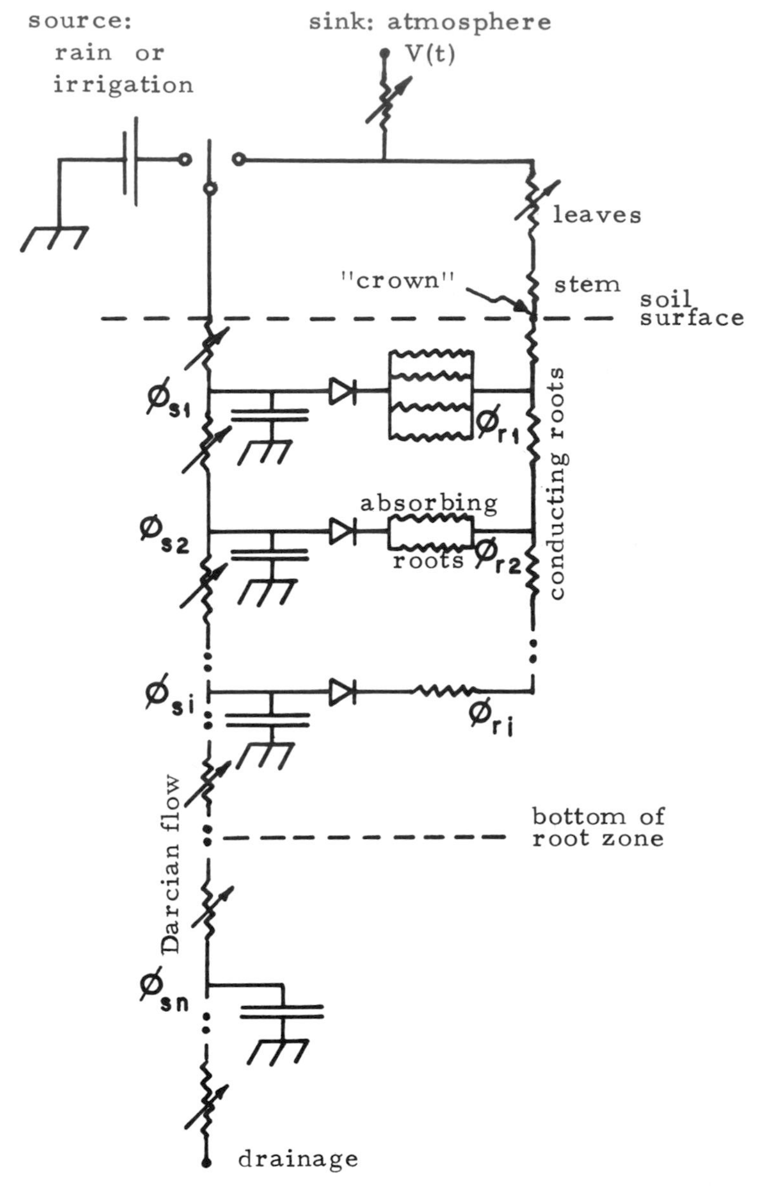

Figure 5.1. Schematic representation of a root system as a resistance network. Soil layers are shown as capacitors, linked by the variable resistance of unsaturated vertical flow and discharged by the roots through the variable resistance of the canopy. The roots are represented by a resistance to absorption and a resistance to conduction (the former inversely proportional to rooting density in each layer, and the latter directly proportional to depth). The diodes at each layer indicate one-directional flow into the roots. The atmospheric sink is shown to be of variable potential. The battery at upper left represents a source of water recharging the soil layers during episodes of rainfall.

the contributions of all volume elements within the root zone:

$$Q = \sum_{i=1}^{n} \frac{(\phi_S)_i - \phi_C}{(R_r)_i + (R_S)_i}$$

Hence,

$$\phi_C = \left[\sum_{i=1}^{n} \frac{(\phi_S)_i}{(R_r)_i + (R_S)_i} - Q \right] \Big/ \sum_{i=1}^{n} \frac{1}{(R_r)_i + (R_S)_i} \qquad (5.13)$$

where n is the total number of volume elements or layers in the rooting zone. Knowing the value of Q (which, as a first approximation for a freely transpiring plant, is equal to the climatically induced transpirational demand) and the values of $(R_r)_i$ (which depends on the rooting-density distribution in the profile as well as on the depth of the particular layer) as well as ϕ_S and R_S for each layer, we can obtain the value of ϕ_C at successive times by a process of iteration. This will be elucidated in the next section.

The scheme of our model, representing the roots as a resistance network discharging a series of capacitors (representing water-charged soil layers) is shown in Figure 5.1.

C. Description of the Computer Model

Like all other models given in this monograph, the root extraction model we shall present in this section was programmed in System/360 CSMP (Figure 5.2). It is based on the model published by Hillel *et al.* 1976.

INITIAL Section

The following constants are given (note: unless otherwise specified, the dimensions are in MKS units):

(1) NJ = number of compartments comprising the profile
(2) A = a coefficient of root absorption
(3) B = the constant of equation (5.11) being an effective root length coefficient
(4) C = root conductance coefficient
(5) RTL = total length of roots under 1 m^2 of field area (M)
(6) RTRSU = hydraulic resistance of roots per unit length (S/M).

(7) DTRDEM = daily transpiration demand (M/day)

(8) D = molecular diffusion coefficient for solutes in bulk water (M^2/S)

(9) DISP = hydrodynamic dispersion coefficient (M)

(10) ERROR, CF, FLPFLP = parameters pertaining to the calculation of crown potential, to be explained subsequently.

The following data are given in tabular form:

(1) TCOM = thickness of compartments (M).

(2) RRL = Relative root length, being the fractional distribution of roots in the various layers of the soil profile. The data given correspond to root counts taken in a field of Rhodes grass (*Chloris guayana*) in Gilat, Israel (Figure 5.3).

(3) ITHETA = Initial volumetric wetness of the soil profile layers.

(4) ICONC = Initial concentration of the soil solution (K mol/M^3).

(5) LABTB = Relation between wetness and tortuosity.

(6) COTB = Volumetric wetness versus hydraulic conductivity (M/S) pertaining to Gilat sandy loam.

(7) SUTB = Volumetric wetness versus matric suction (M), the soil moisture characteristic, assuming no hysteresis.

The following are calculated:

(1) AVTRD = Average transpirational demand (M/S).

(2) AMP = Amplitude of the daily sine-wave of transpirational demand (M/S).

(3) DEPTH and DIST = Depth of each compartment and distance or length of flow path from midlevel of each compartment to the one above it (M).

(4) PRTL = Partial root length in each compartment, as fraction of the total length of roots in the profile (M).

(5) RSRT = Hydraulic resistance of the active roots in each compartment (S), being inversely proportional to the length of roots in the compartment (PRTL). The expression for RSRT consists of two terms: a constant term representing resistance to absorption, and a depth-dependent term representing resistance to conduction. Thus,

RSRT(I) = RTRSU*(A+C*DEPTH(I))/PRTL(I) (5.14)

(6) IVOLW = Initial volume of water in each compartment volume per unit area of the field (M).

(7) IAMS = Initial amount of salt in each compartment per unit area (K mol/M^2).

Figure 5.2. CSMP listing for macroscopic scale model of water uptake by a nonuniform root system and of water and salt movement in the soil profile.

```
TITLE     SOIL MOISTURE EXTRACTION BY NONUNIFORM ROOT SYSTEM
*              INCLUDING THE MOVEMENT OF WATER AND SOLUTES
*                          THROUGH THE SOIL PROFILE

*               UNITS
*            KG = KILOGRAMS
*            KMOL = KILOMOLES
*            M  = METERS
*            S  = SECONDS

*                              GLOSSARY OF SYMBOLS

* A      = ABSORPTION EFFICIENCY COEFFICIENT FOR ROOTS
* AMP    = AMPLITUDE OF DAILY WAVE OF TRANSPIRATIONAL DEMAND (M/S)
* AMS    = AMOUNT OF SALT IN EACH COMPARTMENT (KMOL/M^3)
* AVCOND = AVERAGE HYDRAULIC CONDUCTIVITY FOR FLOW BETWEEN COM-
*          PARTMENTS (M/S)
* AVTRD  = AVERAGE TRANSPIRATIONAL DEMAND (M/S)
* B      = EFFECTIVE ROOT LENGTH COEFFICIENT
* CF     = CORRECTION FACTOR FOR ITERATIVE ESTIMATION OF CROWN
*          POTENTIAL
* CONC   = CONCENTRATION OF THE SOIL SOLUTION (KMOL/M^3)
* COND   = HYDRAULIC CONDUCTIVITY (M/S)
* COTB   = HYDRAULIC CONDUCTIVITY TABLE FOR SOIL WATER
* CRTEX  = CUMULATIVE ROOT EXTRACTION (M)
* D      = MOLECULAR DIFFUSION COEFFICIENT FOR SOLUTES (M^2/S)
* DA     = DIFFUSION-DISPERSION COEFFICIENT FOR SOLUTE FLOW (M^2/S)
* DEPTH  = DEPTH OF MIDPOINT OF COMPARTMENT (M)
* DIF    = RELATIVE DIFFERENCE BETWEEN TOTAL ROOT EXTRACTION AND
*          TRANSPIRATION RATE (DIMENSIONLESS)
* DISP   = HYDRODYNAMIC DISPERSION COEFFICIENT (M)
* DIST   = DISTANCE OF FLOW BETWEEN ADJACENT COMPARTMENTS (M)
* DTRDEM = DAILY TRANSPIRATION DEMAND (M)
* ERROR  = ERROR CRITERION FOR ITERATIVE DETERMINATION OF CROWN
*          POTENTIAL (DIMENSIONLESS)
* FLPFLP = CRITERION FOR CHANGING DIRECTION OF CALCULATION FOR
*          TOTAL ROOT EXTRACTION
* FLS    = FLOW RATE OF SOLUTE (KMOL/SEC/M^2)
* FLW    = FLOW RATE OF WATER (M/S)
* IAMS   = INITIAL AMOUNT OF SALT IN EACH COMPARTMENT (KMOL/M^3)
* ICONC  = INITIAL CONCENTRATION OF THE  SOIL SOLUTION (KMOL/M^3)
* ITHETA = INITIAL VOLUMETRIC WETNESS (M^3/M^3)
* IVOLW  = INITIAL VOLUME OF WATER IN EACH COMPARTMENT (M)
* J      = INDEX OF COMPARTMENTS (ORDINAL NUMBER)
* NFLS   = NET FLOW OF SALT (KMOL/SEC/M^2)
* NFLW   = NET FLOW OF WATER (M/S)
* NJ     = NUMBER OF COMPARTMENTS (COMPRISING THE SOIL PROFILE)
* PEVAP  = POTENTIAL SOIL MOISTURE EVAPORATION RATE (M/S)
* POTH   = HYDRAULIC POTENTIAL HEAD OF SOIL WATER (M)
* POTM   = MATRIC POTENTIAL HEAD OF SOIL WATER (M)
* POTOS  = OSMOTIC POTENTIAL HEAD OF SOIL WATER (M)
* POTRT  = POTENTIAL HEAD OF WATER IN THE ROOTS (M)
* PRTL   = PARTIAL ROOT LENGTH IN EACH COMPARTMENT (M)
* PTOTL  = TOTAL SOIL MOISTURE POTENTIAL HEAD (MATRIC, GRAVITATION-
*          AL, AND OSMOTIC, M)
* RRL    = RELATIVE ROOT LENGTH
* RSRT   = HYDRAULIC RESISTANCE OF THE ROOTS IN EACH COMPARTMENT (S)
* RSSL   = HYDRAULIC RESISTANCE OF SOIL IN EACH COMPARTMENT (S)
* RTEX   = ROOT EXTRACTION RATE OF SOIL MOISTURE FROM EACH COMPART-
*          MENT (M/S)
* RTL    = LENGTH OF ROOTS IN SOIL PROFILE PER 1 M^2 OF FIELD AREA
*          (M)
* RTRSU  = HYDRAULIC RESISTANCE OF ROOTS PER UNIT LENGTH (S/M)
* SUCTB  = SUCTION TABLE FOR SOIL WATER
* TCOM   = THICKNESS OF COMPARTMENT (M)
* THETA  = WETNESS OF SOIL IN EACH COMPARTMENT (VOLUME FRACTION,
*          (M^3/M^3)
* TORTB  = TORTUOSITY TABLE (VOLUMETRIC WETNESS VERSUS TORTUOSITY
*          FACTOR)
* TRDEM  = TRANSPIRATIONAL DEMAND (M/S)
* VOLW   = VOLUME OF WATER IN EACH COMPARTMENT (M)
```

```
STORAGE       TCOM(20),DEPTH(20),RRL(20),PRTL(20),DA(20),ITHETA(20)
STORAGE       ICONC(20),PTOTL(20),RSRT(20),DIST(20),CONC(20),THETA(20)
STORAGE       FLW(20),FLS(20),COND(20),AVCOND(20),POTRT(20),POTH(20)
STORAGE       POTM(20),POTOS(20),RSSL(20)
/       DIMENSION RTEX(20),CRTEX(20),AMS(20),IAMS(20),NFLS(20),VOLW(20)
/       DIMENSION IVOLW(20),NFLW(20)
/       EQUIVALENCE (AMS1,AMS(1)),(IAMS1,IAMS(1)),(NFLS1,NFLS(1)),(RTEX1,RTEX(1))
/       EQUIVALENCE (VOLW1,VOLW(1)),(IVOLW1,IVOLW(1)),(NFLW1,NFLW(1))
FIXED         J,NJ,NJJ

INITIAL

NOSORT
PARAMETER     ERROR=.01,FLPFLP=-1.,CF=.01,D=1.E-9,DISP=.02,A=1.,C=1.
PARAMETER     NJ=18,PI=3.14159,B=1.,RTL=1.E4,RTRSU=1.E7,DTRDEM=.01
              NJJ=NJ+1
              FLS(1)=0.
              AVTRD=DTRDEM/86400.
              AMP+PI*AVTRD
TABLE TCOM(1-18)=15*.05,.10,.20,.30
TABLE RRL(1-18)=2.455E-1,2.2E-1,2.E-1,1.5E-1,1.E-1,.5E-1,.2E-1,.1E-1,...
              .3E-2,.1E-2,.3E-3,.1E-3,.3E-4,.1E-4,.3E-5,.1E-5,.3E-6,.3E-7
TABLE ICONC(1-18)=18*.02
TABLE THEAT(1-18)=18*.25
              DEPTH(1)=.5*TCOM(1)
              DIST(1)=DEPTH(1)
       DO 20 I=2,NJ
              DIST(I)=.5*(TCOM(I-1)+TCOM(I))
           20 DEPTH(I)=DEPTH(I-1)+DIST(I)
       DO 30 I=1,NJ
              PRTL(I)=RTL*RRL(I)
              RSRT(I)=RTRSU*(A+C*DEPTH(I))/PRTL(I)
              IVOLW(I)=ITHETA(I)*TCOM(I)
           30 IAMS(I)=ICONC(I)*IVOLW(I)
FUNCTION COTB=(.01,.788E-13),(05,.278E-12),(0.75,.925E-12),(.1,.138E-11),...
              (.15,2.5E-11),(.2,5.E-10),(.25,5.E-9),(.3,4.E-8),(.35,1.5E-7),...
              (.4,.6E-6),(.45,2.E-6),(.459,2.1E-5),(1.,2.1E-6)
FUNCTION SUTB=(0.1,500.),(.05,200.),(.075,150.),(.1,42.5),(.15,10.8),(.2,6.),...
              (.23,3.2),(.25,3.),(.3,2.2),(.35,1.),(.4,.29),(.459,0.),1.,-60)
FUNCTION LABTB=(.0,.25),(.45,.67)

DYNAMIC

NOSORT
              VOLW1-INTGRL(IVOLW1,NFLW1,18)
              AMS1=INTGRL(IAMS1,NFLS1,18)
      DO 100 I=1,NJ
              THETA(I)=VOLW(I)/TCOM(I)
              CONC(I)=AMS(I)/VOLW(I)
              POTM(I)=-AFGEN(SUTB,THETA(I))
              POTH(I)=POTM(I)-DEPTH(I)
```

```
          COND(I)=AFGEN(COTB,THETA(I))
          RSSL(I)=1./(B*COND(I)*PRTL(I))
          POTOS(I)=-CONC(I)*488.
      100 PTOTL(I)=POTH(I)+POTOS(I)
          TRDEM=AMAX1(.01*AVTRD,.99*AMP*SIN(2.*PI*TIME/86400.))
          IF (TIME.EQ.0.) POTCR=POTH(1)+POTOS(1)
    DO 110 I=2,NJ
          AVCOND(I)=.5*(COND(I-1)+COND(I))
          FLW(I)=AVDOND(I)*(POTH(I-1)-POTH(I))/DIST(I)
          DA(I)=D*.5*(THETA(I-1)+THETA(I))*AFGEN(LABTB,.5*(THETA(I-1)+...
                THETA(I)))+DISP*FLW(I)/(.5*(THETA(I-1)+THETA(I)))
      110 FLS(I)=FLW(I)*.5*(CONC(I-1)+CONC(I))+DA(I)*(CONC(I-1)-CONC(I))/DIST(I)
          PEVAP=.02*TRDEM
          IF (POTM(1).GT.-500.) FLW(1)=PEVAP
          IF (POTM(1).LE.-500.) FLW(1)=FLW(2)
          CEVAP=INTGRL(0.,-FLW(1))
          FLW(NJJ)=COND(NJ)
          FLS(NJJ)=CONC(NJ)*FLW(NJJ)
*     CALCULATION OF POTCR (POTENTIAL OF ROOT CROWN)
          FLPFLP=-FLPFLP
      115 CONTINUE
          SUMR=0.
    DO 150 J=1,NJ
          I=J
          IF (FLPFLP.EQ.1.) I=NJ-J+1
          RTEX(I)=AMAX1(0.,(POTH(I)+POTOS(I)-POTCR)/(RSSL(I)+RSRT(I)))
          IF (SUMR.LE.TRDEM) RTEX(I)=AMIN1(RTEX(I),TRDEM-SUMR)
          SUMR=SUMR+RTEX(I)
      150 CONTINUE
          DIF=(SUMR-TRDEM)/TRDEM
          IF (ABS(DIF).LE.ERROR) GO TO 165
          POTCR=POTCR-(DIF*POTCR*CF)
          GO TO 115
      165 CONTINUE
          CRTEX1=INTGRL(0.,RTEX1,18)
    DO 170 I=1,NJ
      170 POTRT(I)=POTCR+RTEX(I)*RSRT(I)
    DO 120 I=1,NJ
          NFLW(I)=FLW(I)-FLW(I+1)-RTEX(I)
      120 NFLS(I)=FLS(I)-FLS(I+1)

TERMINAL

PRINT (optional)
PRTPLT (optional)
TIMER FINTIM=1296000., OUTDEL=21600.
FINISH POTCR=-300.
END
STOP
ENDJOB
```

DYNAMIC Section

Operations specified in this section are repeated at each timestep during the simulation. The following variables are calculated, as shown in the program (Figure 5.2).

(1) VOLW = The volume of water in each compartment (M).
(2) AMS = Amount of salt in each compartment (K mol/M^2).
(3) THETA = Volumetric wetness.
(4) CONC = Concentration of the soil solution (K mol/M^3).
(5) POTM = Matric potential, obtainable by interpolation of the tabulated soil moisture characteristics in the *INITIAL* section (M).
(6) POTH = Hydraulic potential head (M).
(7) COND = Hydraulic conductivity, obtainable by interpolation of the conductivity versus wetness table (M/S).
(8) RSSL = Hydraulic resistance of the soil, a function of conductivity and length of flow path in each compartment (S). The flow path of water in the soil toward the root is taken to be inversely proportional to the length of roots per unit volume.
(9) POTOS = Osmotic potential of the soil solution, in head units (M).
(10) PTOTL = Total soil moisture potential head (M).
(11) TRDEM = Transpirational demand (M/S), being the positive portion of a sine function of time:

```
TRDEM = AMAX1(0.01*AVTRD,
0.99*AMP*SIN(2.*PI*TIME/86400.)                    (5.15)
```

Note that nighttime transpiration rate is taken to be 1 percent of the average rate of the diurnal period. TRDEM can also be given in tabular form allowing for irregular variations during each day and between days.
(12) AVCOND = Average hydraulic conductivity for flow segments between adjacent compartments (M/S).
(13) FLW = Flow rate of water, by Darcy's Law (M/S).
(14) DA = Combined diffusion-dispersion coefficient (M^2/S).
(15) FLS = Flow rate of salt, by convection and diffusion-dispersion (K mol/S per M^2).
(16) PEVAP = Potential rate of direct evaporation of soil moisture (M/S). The actual rate of evaporation is either equal to the potential rate, or determined by the rate at which the profile delivers water to the dried surface zone. This is specified in the IF statements following statement 110 in the program.
(17) CEVAP = The cumulative evaporation.
(18) FLW(NJJ) and FLS(NJJ) = The flow rates for water and solutes, respectively, across the bottom boundary to the soil profile. Unit hydraulic gradient is herein assumed to operate at that depth (which is considerably beyond the rooting zone) at all times. Alternative bottom boundary conditions can be substituted as appropriate.

Calculation of crown potential and root extraction rates

The total transpirational demand is given above in terms of the time-variable TRDEM [equation (5.15)].

To answer this demand, the roots extract water from each layer according to:

$$RTEX(I) = \frac{PTOTL(I)-POTCR}{RSSL(I)+RSRT(I)} \tag{5.16}$$

where POTCR, the crown potential, is unknown. We assume that the plant generates a POTCR value just sufficient to satisfy TRDEM as the summation of all the individual root extraction terms $\Sigma RTEX(I)$.

The search for the appropriate POTCR value at any point in time is carried out according to the following algorithm:

(1) Compute TRDEM as in Equation (5.14).

(2) Compute RSRT(I) for each compartment as shown in the INITIAL section, as well as PTOTL(I) and RSSL(I) in the DYNAMIC section, as shown above.

(3) Compute RTEX(I) by Equation (5.16), using the current value of POTCR. [Initially, POTCR is set equal to PTOTL(1).]

(4) Sum up the total extraction

$$SUMR = \sum_{I=1}^{m} RTEX(I), \; m \leq n \tag{5.17}$$

where m is the number of compartments containing roots, and n is the total number of compartments comprising the profile.

(5) Calculate the relative difference (DIF) between the summed root extraction (SUMR) and the actual transpiration rate (TRDEM)

$$DIF = (SUMR-TRDEM)/TRDEM \tag{5.18}$$

(6) If the absolute value of this relative difference is less than or equal to an arbitrarily small error, then POTCR is taken to be equal to its current value and the search procedure is ended:

```
IF (ABS(DIF).LE.ERROR) GO TO XXX                    (5.19)
```

where XXX is the branching number in the main program marking the point of return after using the iterative subroutine for crown potential.

(7) Otherwise, adjust POTCR as follows:

```
POTCR = POTCR-DIF*POTCR*CF                          (5.20)
```

where the correction factor CF is a positive quantity, small enough to prevent inordinate errors yet large enough to avoid the need for an excessive number of iterations. (In the example given, CF = 0.01). After adjustment of POTCR, the procedure returns to step (3) and thence through (4) and (5) to (6), until the criterion for ending the search is satisfied.

The summation process of step (4), Equation (5.17), is conducted between I = 1 to m, which runs alternately from the top layer downward and from the bottom layer upward (that is, from I = NJ - J for J = 1, 2, ..., (NJ - 1)). This is achieved by means of the FLPFLP (DO 150) procedure, in order to prevent the accumulation of errors which might result from inequitable extraction of water from either top or bottom layers.

The remainder of the program deals with output and is mainly formal. The TERMINAL section defines calculations which are to be executed after FINTIM (end of the simulation) has been reached and the method of presenting the data.

D. Results for a Stable, Non-Uniform Root System

The results of a series of simulation runs for non-uniform but stable root systems are shown in Figures 5.4 to 5.16.

A preliminary "basic" simulation run was carried out to establish a general pattern of water extraction as a basis for later comparisons of variable effects. The following conditions were assumed to prevail:

(1) Total root length (RTL) = 10,000 m of roots per m^2 of field.
(2) Transpirational demand: Sinusoidal (diurnally fluctuating), totaling 10 mm per day.
(3) Evaporation rate for soil surface = 2 percent of transpiration rate.
(4) Initial soil moisture content (volumetric wetness) = 25 percent.
(5) Initial concentration of the soil solution = 0.02 normality.
(6) Root distribution pattern: as shown in the top curve of Figure 5.3.

Figure 5.4 presents successive soil moisture profiles during the process of extraction from a soil initially at 25 percent moisture, which is roughly the "field capacity" for Gilat silt loam. Note that 99.955 percent of the roots are present in the upper half-metre of the soil profile (upper curve, Figure 5.3). An interesting feature of this family of curves is that only the top 30 cm or so of the profile exhibits anything like a

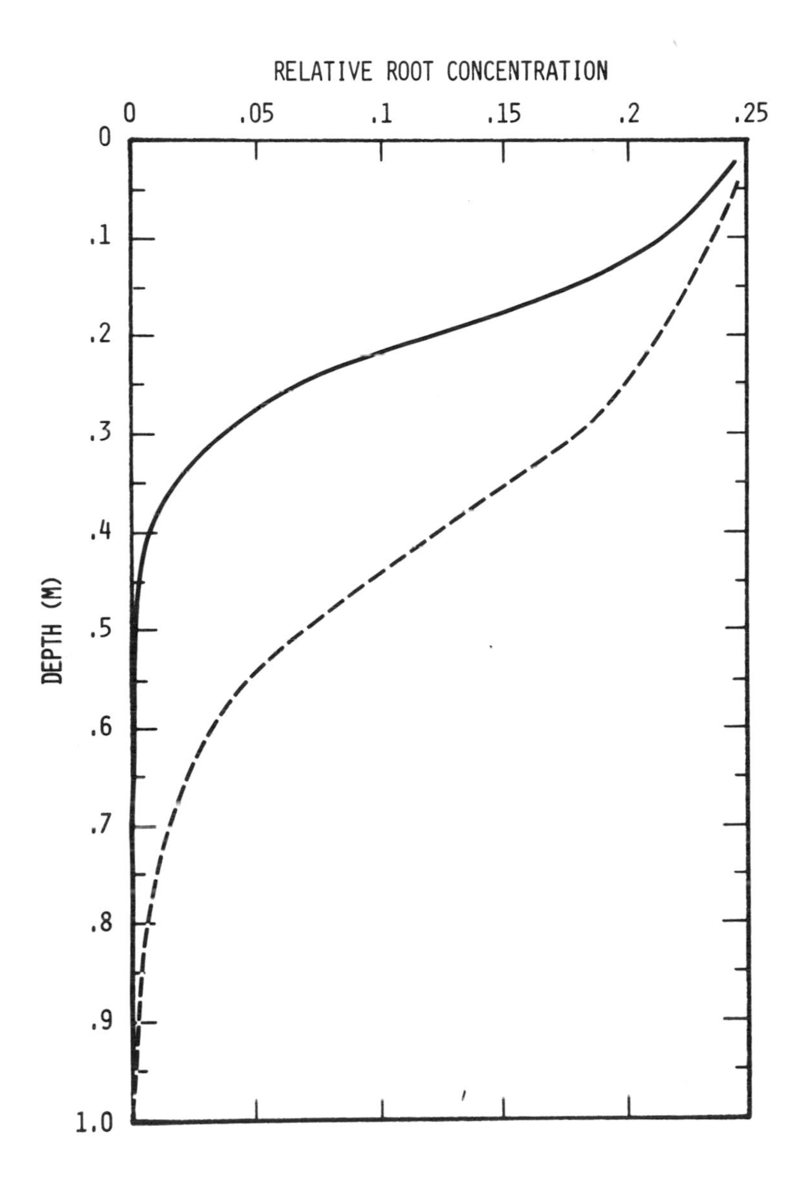

Figure 5.3. Relative root distribution in the soil profile: Solid curve: basic simulation run, "shallow root system"; dashed curve: "deep root system."

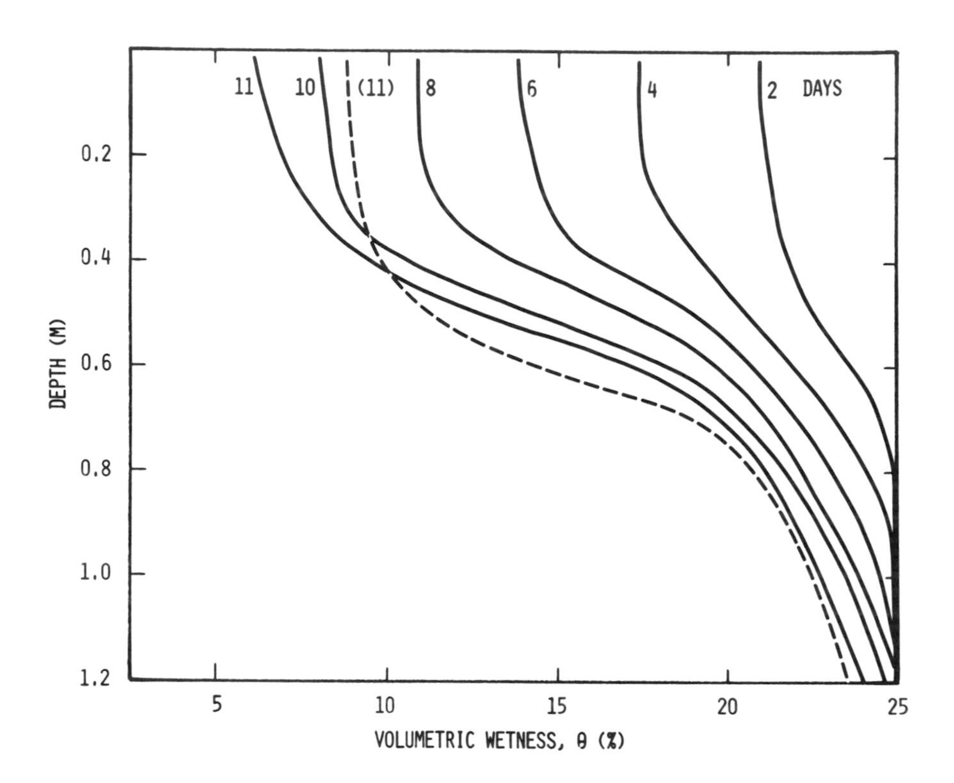

Figure 5.4. Basic simulation run: Pattern of soil moisture extraction by roots as shown by successive wetness profiles. The numbers alongside the curves indicate days from time zero, at which soil wetness was a uniform 25 percent. The dashed curve is for a root system 10-times as dense (RTL = 10^5 m) having the same relative distribution in the profile.

uniform value of wetness at any particular time, whereas a considerable zone below is highly nonuniform.

The pattern of moisture extraction is characterized by three discernible phenomena: Gradual reduction of wetness within the zone of major root concentration, gradually deepening zone of moisture extraction, and gradually steepening gradients of moisture between the untapped subsoil and the zone of major extraction. On the eleventh day, at which our hypothetical plant is presumed to be in a state of stress leading to wilting (the crown potential having fallen below -30 bars) the moisture profile varied from about 6 percent in the top 20 cm to about 10 percent at 40 cm, about 17 percent at 60 cm, and over 20 percent below 80 cm.

The highly variable nature of the soil moisture profile is also illustrated in Figure 5.5, which presents successive matric potential profiles. The concurrently developing profile of total soil moisture potential is shown for the 11th day, at which wilting was presumed to have occurred. It is seen that the soil moisture potential varies more and more widely within and below the root zone as the process of water extraction progresses. On the 11th day, the matric suction varied from less than 2 bars at the 50 cm depth to more than 17 in the upper 30 cm.

The actual pattern of cumulative extraction by roots is shown in Figure 5.6. It is seen that with time, as the topmost layer of soil is progressively depleted by both root extraction and direct evaporation, more and more of the plant's water supply is extracted from the deeper layers even though the roots are very sparse beyond, say, 45 cm. This is apparently a consequence of the considerable flux of water transmitted to the drying root zone from the wetter subsoil layers.

The transmission of water through the 75-cm plane (taken to be the bottom of the root zone) is illustrated in Figure 5.7. Downward drainage is seen to take place, albeit at a diminishing rate, during the first 3 days, after which capillary rise begins and fluctuates in a repetitive diurnal pattern. The rate of capillary rise begins to diminish toward the end of the simulation period. The total drainage during the first 3 days amounted to only 1 mm, while the total capillary rise from the third to the eleventh day amounted to nearly 9 mm (*i.e.*, about 1.5 mm per day, or 15 percent of the transpirational demand).

The transmission of salt through the bottom of the root zone is shown in Figure 5.8. The downward and upward movement of salt is seen to be concurrent with the flux of water during the first 9 days or so, as the main mechanism of salt transport is by convection. Gradually, however, as the concentration in the soil solution within the root zone increases (through evaporation and the selective extraction of water by the roots), concentration

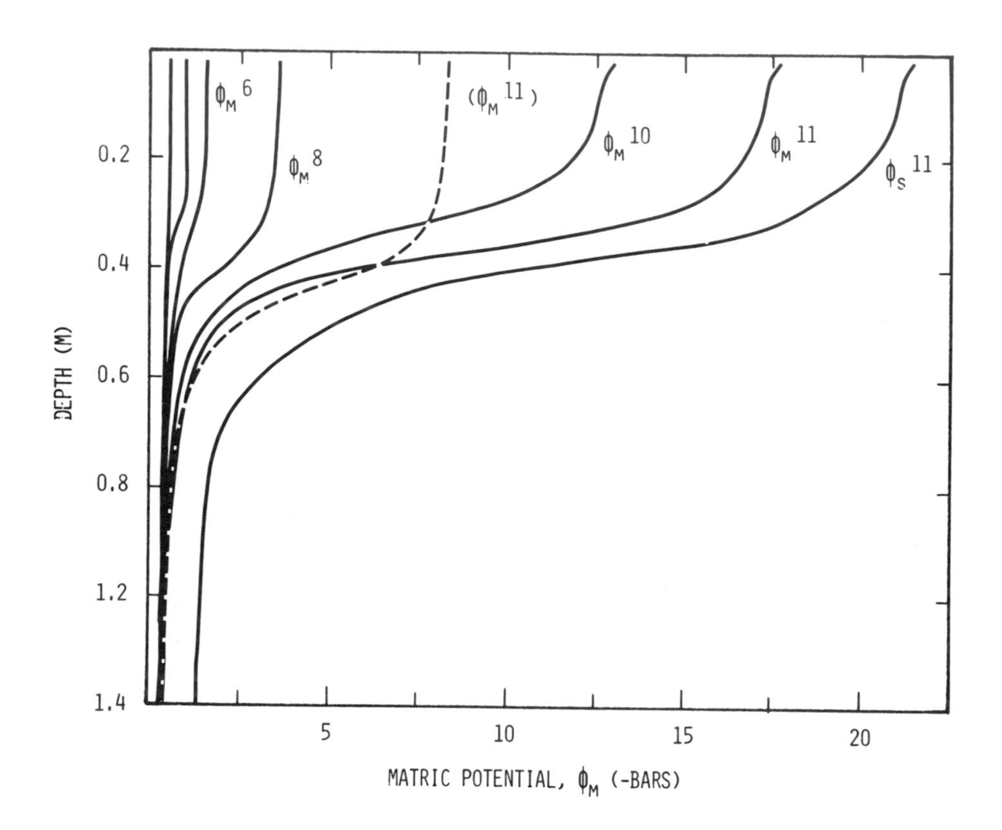

Figure 5.5. Basic simulation: Successive profiles of matric potential (ϕm) on days 2, 4, 6, 8, 10, and 11; and of total water potential (ϕs) on day 11. The dashed curve is for the dense root system.

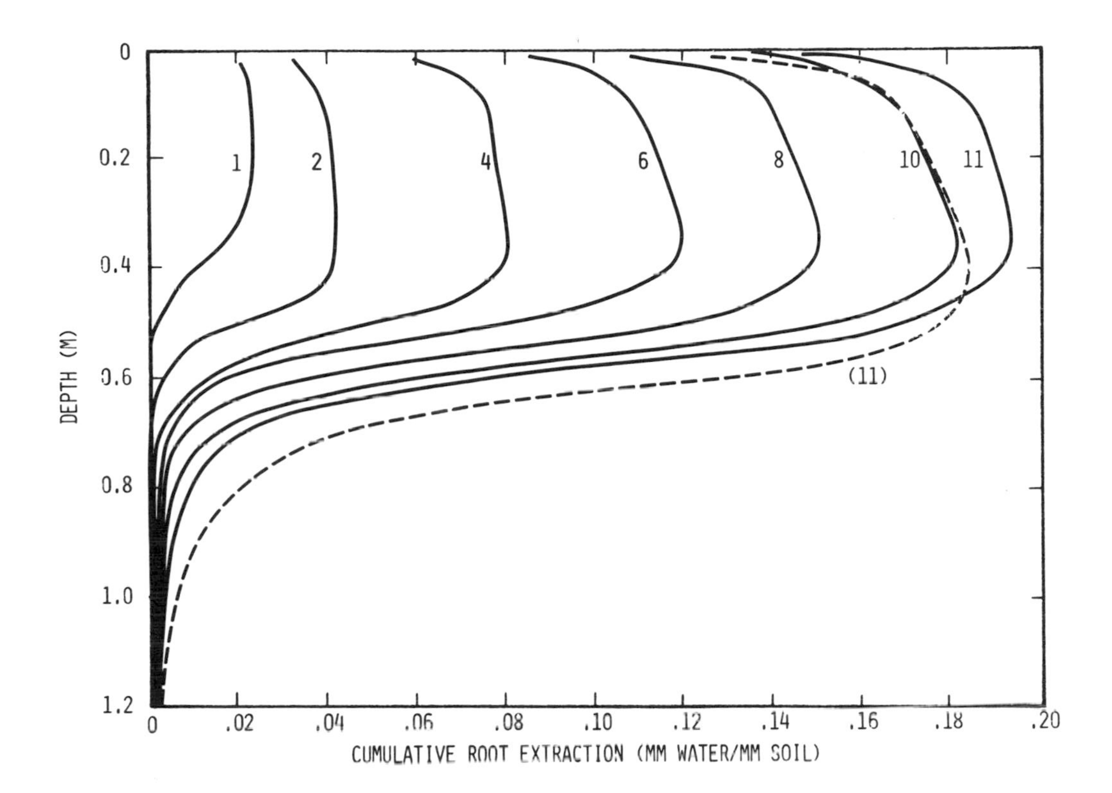

Figure 5.6. "Basic" simulation: Cumulative root extraction of soil moisture as a function of depth at different days during the 11-day simulation. The dashed curve is for the denser root system.

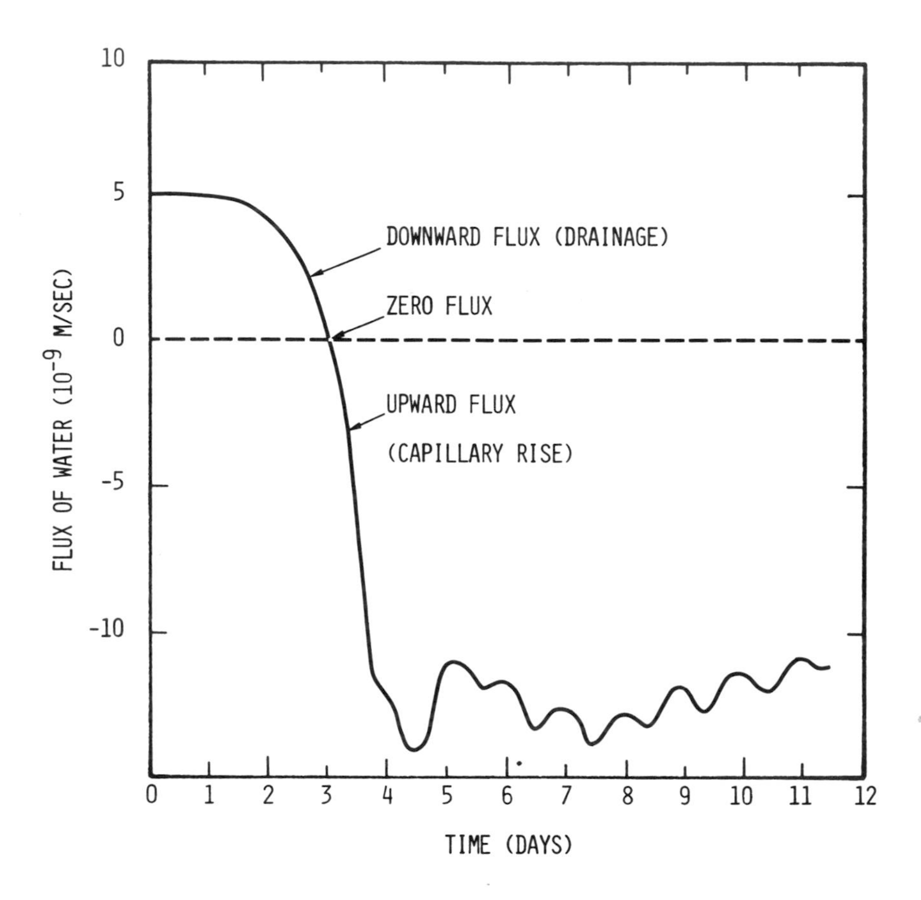

Figure 5.7. Flux of water through the bottom of the root zone (75 cm depth) as function of time. Basic simulation run.

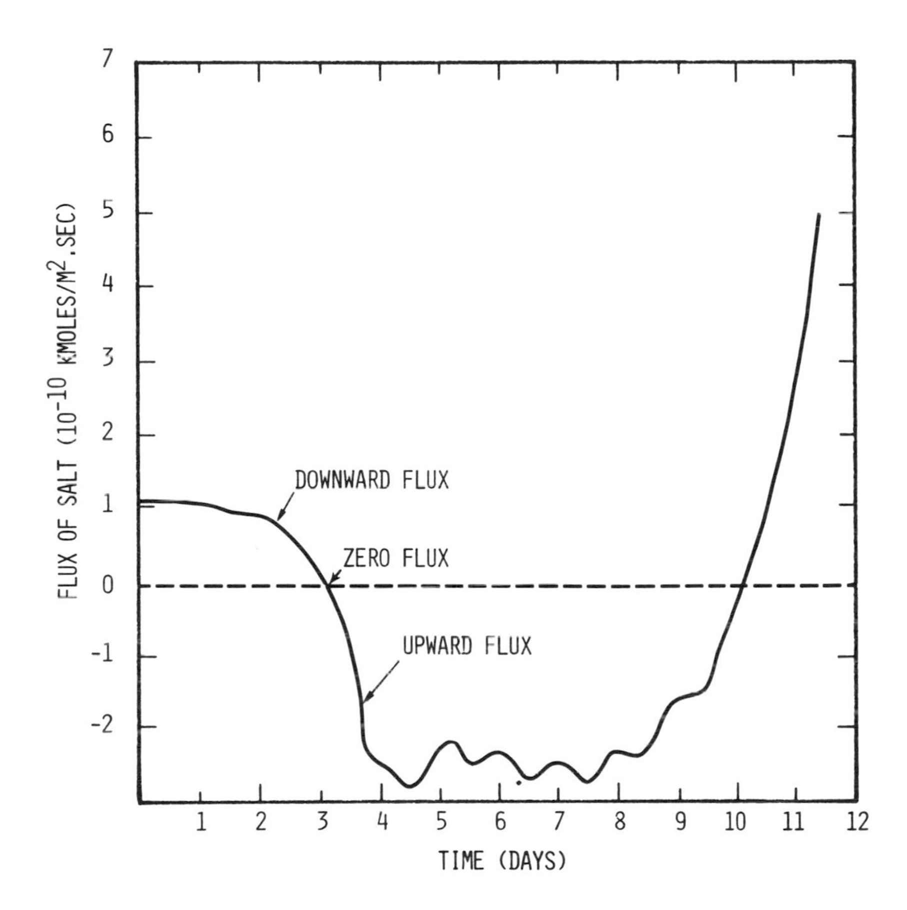

Figure 5.8. Flux of salt through the bottom of the root zone (75 cm depth) as function of time. Basic simulation run.

gradients build up between the root zone and the subsoil, causing downward diffusion at a rate which eventually exceeds the upward transport of salts carried by the convective stream of water.

Figure 5.9 shows the continuous and steepening change in water potential at different segments of the system, in the soil and in the plant. At a depth of about 30 cm, the soil's matric suction, although initially lower than the osmotic, eventually becomes far greater. On the other hand, at a depth of about 50 cm the matric suction remains lower than the osmotic, which builds up on the lower horizons as a consequence of the influx of water and solutes into the lower part of the root zone, where more and more of the water is taken up. (Note that we use the term "suction" to express the negative water potential as a positive quantity).

Figure 5.10 shows the change in crown potential in relation to the change in total soil moisture potential (hydraulic and osmotic) at different depths within the root zone. It is seen that the plant water potential must exceed and increasingly diverge from the soil moisture potential of all depths (particularly the lower depths, where the roots are sparse) for the plant to maintain its transpiration rate in the face of diminishing soil moisture potential which is associated with a steep decrease of hydraulic conductivity.

Following the basic simulation run thus far described, a number of factors were varied to map out their possible effects. Of the numerous conceivable comparisons the model enables us to make, we have chosen but a few for purposes of illustration.

The effect of initial soil wetness is shown in Figure 5.11, in which the time-course of plant water potential is followed during a continuous period of evaporation without replenishment of soil moisture by rain or irrigation. Initial post-irrigation soil wetness would normally depend primarily upon soil texture and profile layers, and to a lesser degree upon quantity and mode of irrigation. In our case, the comparison between initial wetness values of 15, 25, and 35 percent is entirely hypothetical, as it is made for fully and uniformly wetted profiles of the same soil. It is seen that plant water potential fell steeply and rapidly to reach a stressful level (below -30 bars) within only 5 days where the initial wetness was 15 percent. The same hypothetical plant, subject to the same transpirational demand, exhibited a very gradual change of water potential, which reached only -7 bars on the fifteenth day, where the initial soil wetness was 35 percent. The intermediate initial wetness level of 25 percent representing the approximate value of this soil's "field capacity," allowed about 12 days before the stressful condition developed.

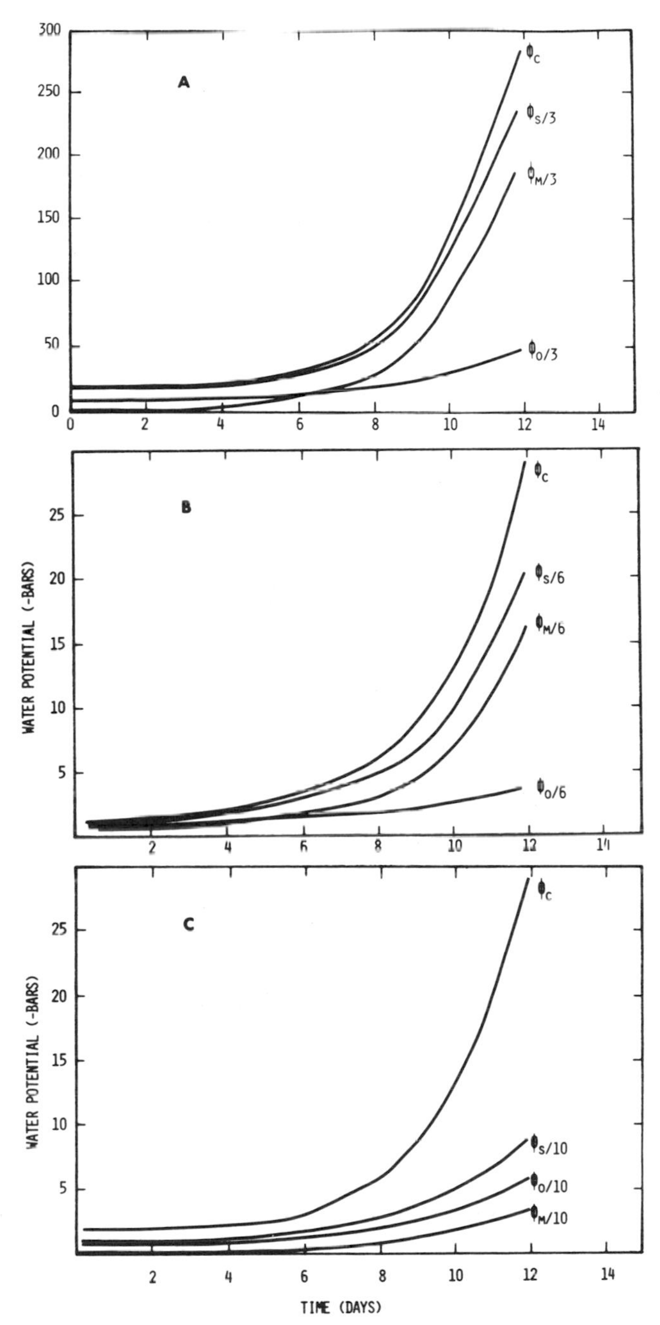

Figure 5.9. Changes in water potential in the soil and plant during the extraction process: ϕ_C = crown potential; $\phi_{S/3}$, $\phi_{S/6}$, $\phi_{S/10}$ = total soil water potential at 15, 30, and 50 cm depths respectively; $\phi_{m/3}$, $\phi_{m/6}$, $\phi_{m/10}$ = soil matric potential at the same depths; $\phi_{O/3}$, $\phi_{O/6}$ $\phi_{O/10}$ = osmotic potential of soil water at the same depths. Basic simulation run. A, B, and C represent the 3rd, 6th and 10th day, respectively.

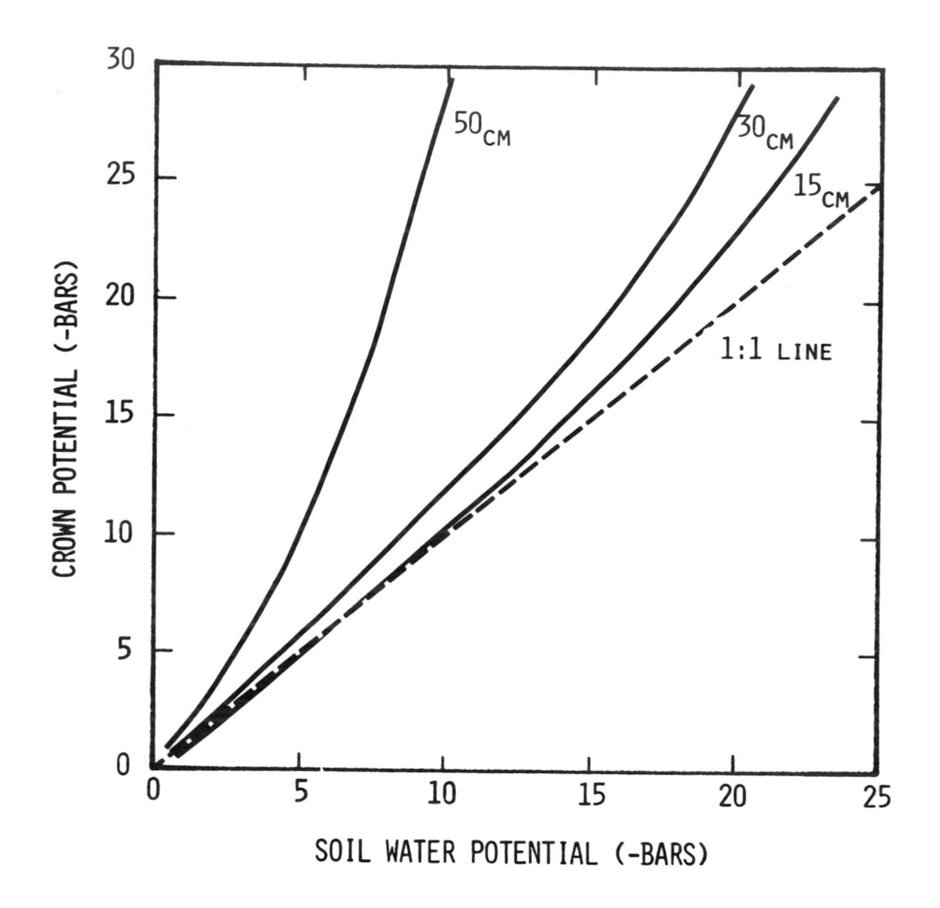

Figure 5.10. Crown potential as function of soil water potential at different depths (15, 30, and 50 cm) within the root zone. Basic simulation run.

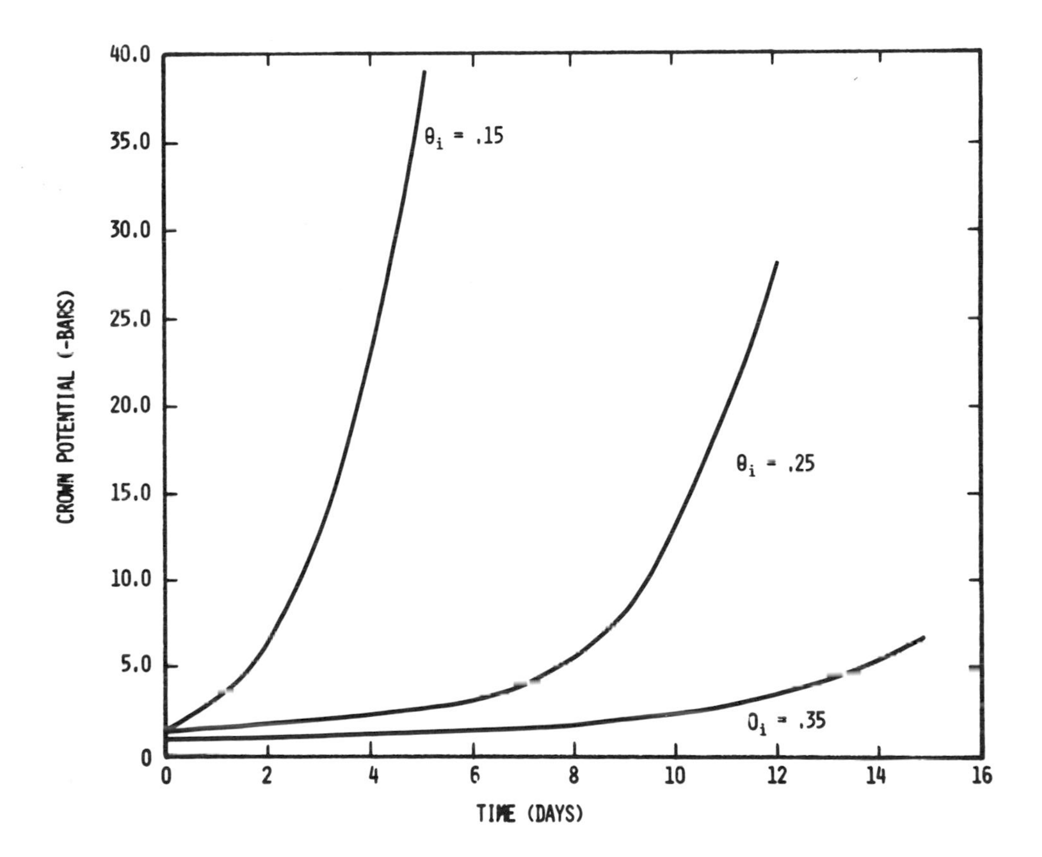

Figure 5.11. Effect of initial soil wetness on the noontime values of plant water potential during a succession of daily transpiration cycles without replenishment of soil moisture.

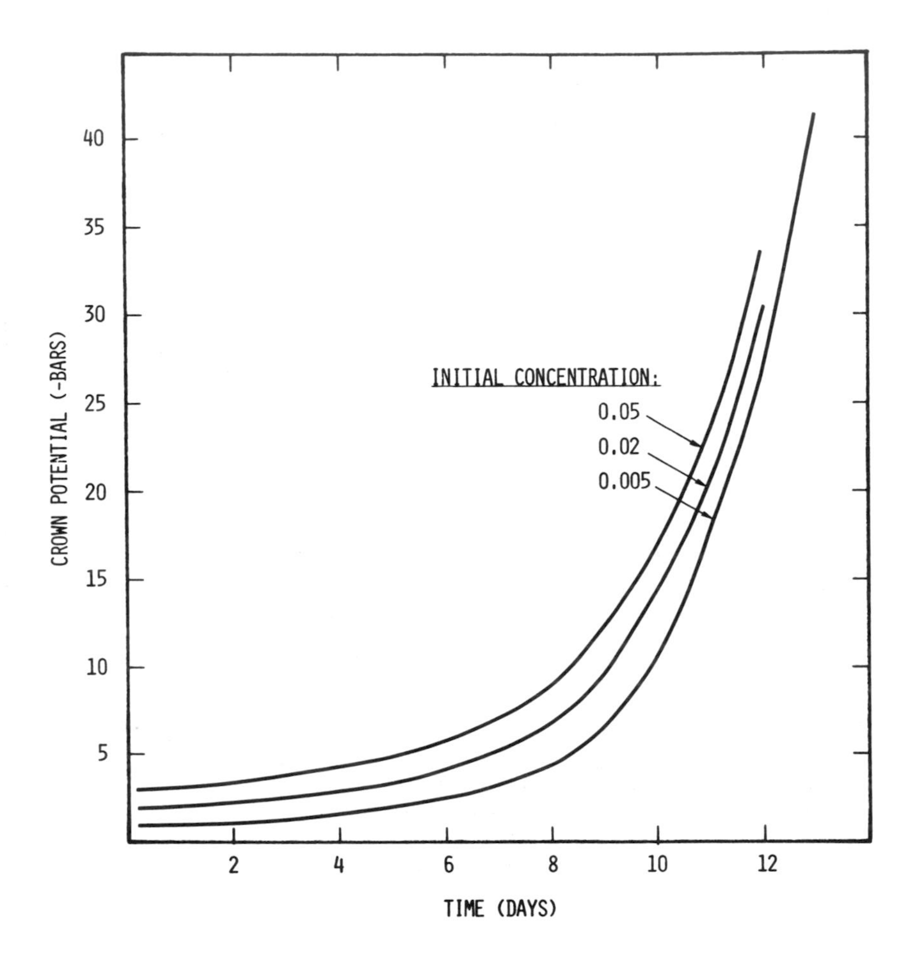

Figure 5.12. Effect of initial concentration of the soil solution on the noontime values of plant water potential during a succession of daily transpiration cycles. The concentrations are given in terms of moles/litre.

Initial concentration of the soil solution apparently affects the development of crown potential in a similar way, though to a lesser degree in the range simulated. This is shown in Figure 5.12. The highest initial concentration tested was of a 0.05 normal solution having an osmotic potential of about -2.5 bars. This is presumably equivalent to irrigation water with nearly 3000 ppm of sodium chloride. This initial concentration caused a steeper and earlier decrease of plant water potential than in the case of the initially less concentrated soil solution. The least concentrated soil solution (5 milliequivalents per litre, with an osmotic potential of -0.25 bar) allowed the plant more time before a stressful condition developed, but this time-increase amounted to only one day under the conditions of our simulation.

The effect of rooting density is shown in Figure 5.13. Here two rooting systems are compared, having a total length of active roots amounting to 100,000 versus 10,000 metres per square metre of field. Both root systems (termed "dense" and "sparse," respectively) were otherwise similar in depth and in distribution within the soil profile. With the sparse root system, our hypothetical plant presumably developed a stressful condition on the 12th day whereas the dense root system allowed it to remain under the stress threshold for nearly 3 days longer in the same soil and climate. Our assumed stress threshold of -30 bars crown potential is of course arbitrary, but the choice of any other value would in principle indicate a similar result.

Figure 5.13 also provides a comparison between a typically fluctuating diurnal evaporative regime and a hypothetical steady one. The latter is obviously an unrealistic condition, and was included in our comparisons only because it had been assumed in some earlier analyses of soil-plant-atmosphere interrelations. In principle, the same trend of a steepening drop in plant water potential, necessary to continue satisfying the transpirational demand, is indicated for both the steady and diurnally fluctuating evaporative regimes. However, the former regime avoids the stress-inducing noontime peaks in transpirational demand and hence allows the plant at least a day longer before a stressful condition is reached.

The decisive influence of rooting depth on the pattern of plant-soil-water relations is illustrated in Figure 5.14. Here, the root system represented by the top curve of Figure 5.3 was compared with one that is twice as deep but otherwise equal in total root length and relative distribution. To simulate the deeper root system, the compartments comprising the profile were simply taken to be twice as thick. The total moisture reserve in the root zone thus was made twice as great. Consequently, the length of time the plant could remain above any stress level was approximately doubled.

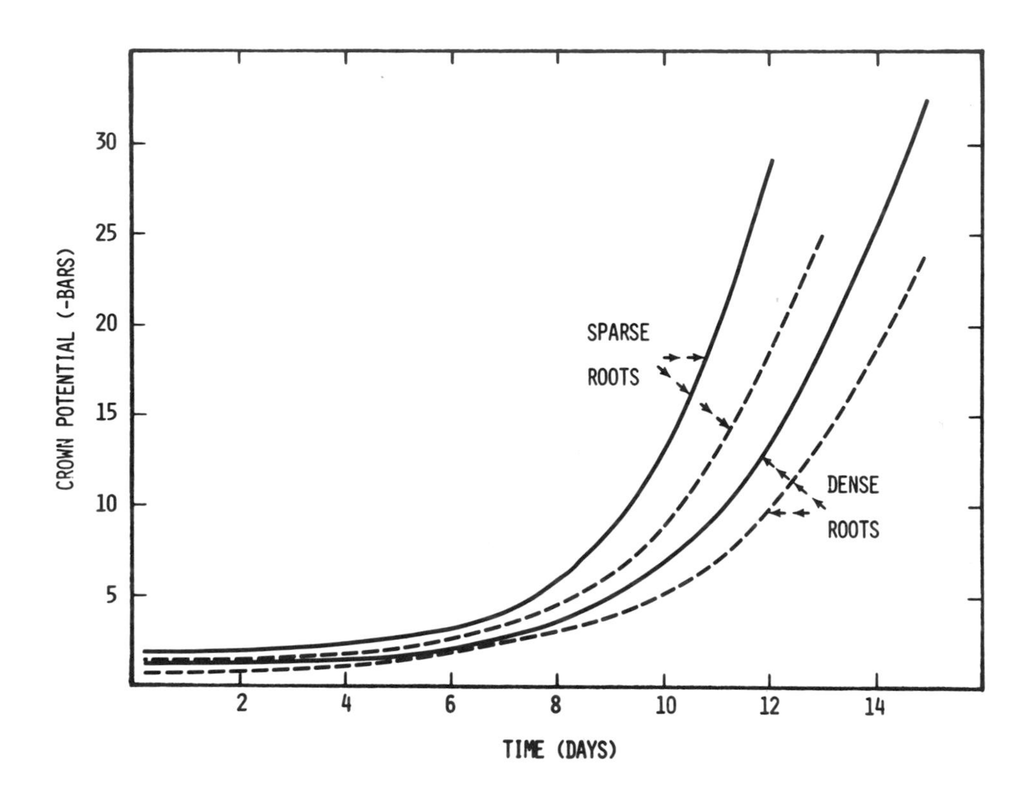

Figure 5.13. Effect of rooting density on the noontime values of plant-water potential during a succession of daily transpiration cycles. "Dense" roots: RTL = $10^5 m/m^2$; "Sparse" roots: RTL = $10^4 m/m^2$. The dashed curves are for a hypothetically steady transpirational demand.

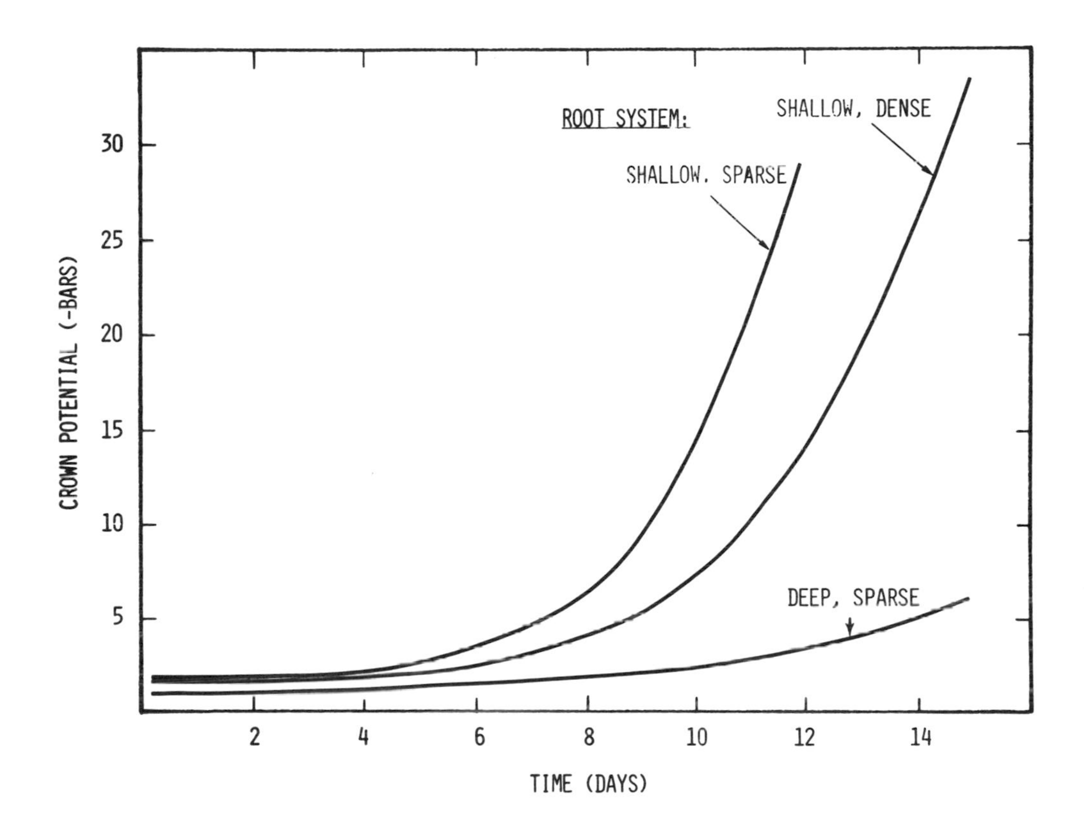

Figure 5.14. Effect of rooting depth on the noontime values of plant-water potential during a succession of daily transpiration waves. "Shallow root system" — as illustrated in Figure 5.2. "Deep root system" — twice as deep, but with equal total root length and relative distribution.

Figures 5.15 and 5.16 indicate the dependence of plant water potential on the magnitude of the transpirational demand. As can be seen in Figure 5.15, an increase of transpirational demand causes an approximately proportional decrease in the length of time the plant can continue to thrive before experiencing any level of water stress. Thus, under a transpirational demand of 20 mm/day, our hypothetical plant underwent a sharp decrease of water potential presumably leading to wilting as early as the sixth day, whereas under a transpiration rate of 10 mm/day it could continue to the twelfth day. Figure 5.16 is a snapshot view of the profiles of matric potential at the end of the fifth successive day of transpiration under three rates of daily transpiration. It is seen that the increase in matric suction at every level of the profile is disproportionate to the rate of transpiration. This disproportionality is a consequence of the nonlinear nature of the soil moisture characteristic which in our model is embodied in the suction table.

Finally, we tested the possible effect of varying the root resistance term. This is illustrated in Figure 5.17, which shows the time course of crown potential with two values of root resistance per unit length. Note that these two values differ by two orders of magnitude, the value we used in our basic simulation run being intermediate between the two. Increasing root resistance is seen to induce a somewhat quicker and steeper decrease of crown potential, but the difference is hardly significant for the conditions of our model, in which the soil resistance term is predominant most of the time, and increasingly so as the soil dries.

Discussion

Certain very definite limitations of the model, as presented, must be borne in mind when its results are evaluated.

Perhaps the most serious omission is that of the stomatal control mechanisms by which the plant can restrict its transpiration rate during periods of high demand as it begins to experience stress. By this mechanism the plant may modify the pattern by which its own water potential responds to the continuous transpirational demand and the dwindling supply of soil moisture. Moreover, the characterization of plant water potential in terms of a single value (namely, the "crown" potential) disregards the potential distribution through the plant system and the possibility that the water potential in the leaves may differ from that in the stem.

Still another shortcoming of our model is its portrayal of the root system as a fixed resistance network allowing no growth or change in conductive properties.

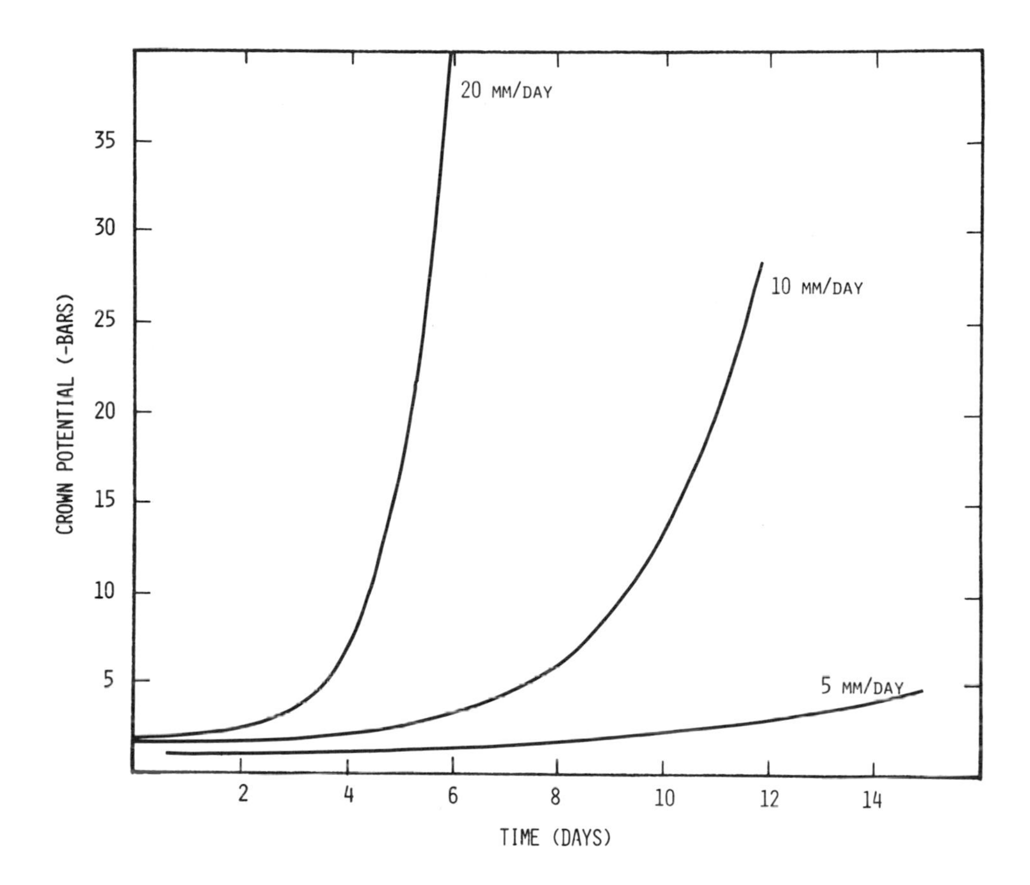

Figure 5.15. Dependence of plant water potential on the transpirational demand.

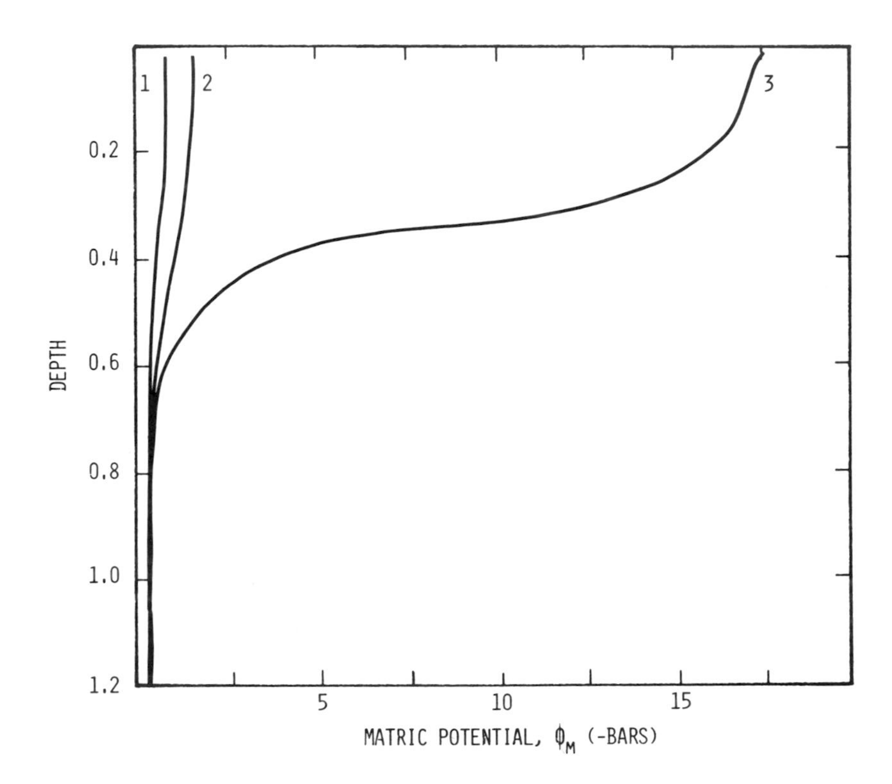

Figure 5.16. Matric potential profiles at the end of 5 days under different transpirational demands: (1) 5 mm/day; (2) 10 mm/day; (3) 20 mm/day.

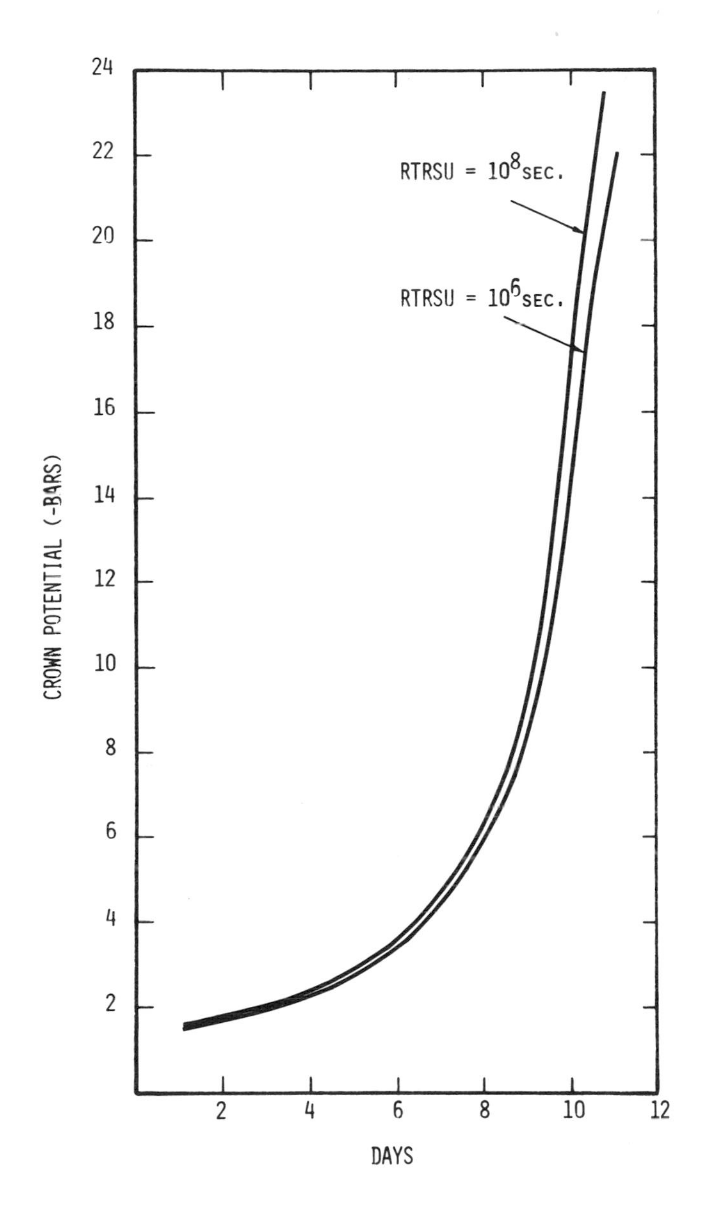

Figure 5.17. Time course of crown potential affected by root resistance.

Finally, our macroscopic scale model disregards the microscopic scale drawdown of soil moisture potential and other phenomena which might occur in the immediate vicinity of each active root (see Hillel *et al.*, 1975b). Such localized effects may be offset at least partly by the ability of roots to grow and thus continuously reach into moist regions of the soil.

As against these limitations, we might point out the model's capability to provide a mechanistic description of the flow of water and salts in and through any soil profile with any distribution of roots under various and variable climatic regimes. The model is flexible enough to allow numerous variations and adaptations in addition to those shown. Though not entirely demonstrated in the necessarily limited range of examples given, the model can readily be adapted to handle variable-intensity rainfall (*e.g.*, Hillel *et al.*, 1975a) or irrigation with water of varying salt concentration, as well as the soil energy balance and thermal regime (*e.g.*, van Bavel and Hillel 1975). Moreover, the model can be made to include water-table conditions or cases of nonuniform soil profiles simply by assigning different suction and conductivity functions, as well as different initial wetness and concentration values, to different layers at various depths.

Many of the model's shortcomings can in fact be corrected by combining it with an appropriate canopy model or by adding subroutines to describe such phenomena as root growth or dependence of root resistance on flux or potential. The difficulty here is that we still seem to lack sufficient knowledge to be able to formulate some of the processes we perceive qualitatively in quantitative mechanistic terms.

A particularly glaring gap in our present knowledge relates to the hydraulics of root systems. Such variables as total vs. effective root length per unit volume of soil, permeability or resistance to absorption and to conduction (and their possible dependence on environmental and physiological factors) are exceedingly difficult to measure. Hence it is difficult to know whether the parameters used in our model, or their numerical values, are realistic or not. Such data as are available in the literature (*e.g.*, Brouwer 1965; Emerson 1954; Wind 1955; Cowan and Milthorpe 1968; Newman 1969; Leyser and Loch 1972) were variously obtained and encompass widely divergent values. Recently, however, studies of roots functioning in fairly natural conditions promise to provide us with some of the desired data (Taylor *et al.* 1972).

Beyond our need for sound and realistic values for specific parameters, we need comprehensive data to characterize the overall performance of root systems in the field. Without such comprehensive data, we do not have truly independent and objective criteria by which to evaluate root extraction models.

Our limited and specific objective in this exercise was to provide a tool to map out systematically some of the basic interactions involved in soil-plant-water relations. For instance, our calculations show that increasing soil moisture content works in the same direction, though not necessarily in a functionally equivalent way, as decreasing transpirational demand, increasing depth and density of rooting, and decreasing salt concentration. Our calculations also show that flow of water through the bottom of the root zone can be a very important component of the water balance, and more specifically that upward capillary flow from the subsoil can contribute significantly to the supply of water to the roots even in the absence of a high water table. Our results also suggest that the leaching of salts out of the bottom of the root zone is rather significantly aided by diffusion processes.

An important lesson to be learned from our results is the degree of approximation involved in the old but still popular attempt to characterize the water status of the root zone in terms of some fixed quantity of "available" or "extractable" soil moisture. The highly variable profiles of moisture content and potential, within and particularly below the root zone, make the choice of where in the profile to measure or how to integrate soil wetness, suction, and salinity over space and time a moot question indeed. The power of the simulation approach is that it can provide an essentially continuous monitoring of the entire system as it varies in response to any number of factors on the basis of cause-and-effect mechanisms.

E. Modification of the Model to Account for Root Growth

The preceding section of this chapter dealt with water uptake by a spacially nonuniform root system but made no provision for the possibility that the configuration of the roots may change in time. A modification will now be presented, based on the work of Hillel and Talpaz (1976), to account for root growth. A plant with growing roots can reach continuously into moist regions of the soil rather than depend entirely on the conduction of water over appreciable distances in the soil against a steadily increasing hydraulic resistance, as is the case with a fixed root system. The process of root growth, if rapid enough, can reduce the effect of the localized drawdown of both matric and osmotic potential around each root, as well as increase the effective volume of soil tapped by the root system as a whole.

While we do not as yet have sufficient information on the hydraulics of growing root systems, the limited data already available (*e.g.*, Taylor and Klepper 1975) suggest that the distribution of roots in the profile can change rather markedly within a period of weeks or even days, particularly in the case of an annual crop. It is therefore of interest to attempt to devise a logical framework for the dynamic simulation of root growth within the context of an overall model of soil water extraction by variously distributed root systems. Such a model might thus serve as a better criterion for the evaluation of soil moisture availability to different types of plants at different stages of growth.

In principle, and insofar as it relates to water uptake, we can consider overall root growth as consisting of several concurrent processes, including proliferation, extension, senescence, and death. As used in the present context, the term proliferation applies to the localized increase of rooting density (*i.e.*, by branching) within each layer without any increase in the volume of the root zone as a whole. Extension is the additional process by which roots from any layer extend downward so as to invade an underlying layer and increase its rooting density. The process of senescence involves suberization and the gradual reduction of root permeability. With further aging, the older roots eventually become totally inactive and, to all intents and purposes, can be considered dead.

In an effort to formulate some of these phenomena in terms compatible with our earlier model, we may write the following equation:

$$R_i^j = R_i^{j-1} + R_i^{j-1}P\Delta t - R_i^{j-1}D\Delta t + R_{i-1}^{j-1}E\Delta t \qquad (5.18)$$

wherein R_i^j is the density of active roots (length per unit volume or number per unit area) in layer i at time j; R_i^{j-1} is root density in the same layer at a previous time, j-1, Δt time units earlier; P, the proliferation rate, is the number of new roots formed per unit time as a fraction of the existing number of roots; D is the death rate per unit time as a fraction of the number of roots present; and E is the extension rate per unit time as a fraction of the number of roots present at the previous, (j-1)th, time step in the overlying layer (i-1).

This formulation disregards the process of senescence and the gradual loss of root absorptivity and conductivity which it might entail. However, even this admittedly simplified representation of root growth requires knowledge of three parameters (namely, P, D, and E), none of which is available to us at present. For want of any proven formulation of root growth dynamics, we tentatively offer a method of modeling root growth based on equation 5.18. In so doing we do not pretend that this is necessarily a realistic formulation, only that it permits, at the present stage of our knowledge, testing the relative sensitivity of the soil-plant-water system to each of the conjectured root growth parameters.

Root Growth Submodel

An algorithm based on equation (5.18) was programmed as a subroutine in the context of the macroscopic scale model of water uptake by a nonuniform root system, presented in the previous sections of this chapter. This subroutine is placed in the DYNAMIC segment of the program, immediately following statement no. 100 and prior to the computation of TRDEM (see Figure 5.2).

We begin by computing the average matric potential of each profile layer for each day of the simulation period.

```
          HIMP = IMPULS(3600.,3600.)
          DIMP = IMPULS(108000.,86400.)
          IF (HIMP.NE.1)GO TO 2000
          DO 1005 I = 1,NJ
     1005 AVPOT(I) = AVPOT(I)+POTM(I)
     2000 CONTINUE
          IF (DIMP.NE.1.)GO TO 3000
          DO 1010 I = 1,NJ
          AVPOT(I) = AVPOT(I)/24.
```

This procedure instructs the computer to read the matric potential value, MPOT(I) in each layer every hour (3600 sec) and to sum these values for each day 86400 sec), starting at noon of the second day (108000 sec). These daily sums of hourly values are then divided by 24 to obtain the mean value of matric potential for the previous day, which is later used as a criterion for determining the proliferation and extension rates for roots in each layer.

We continue within the same DO-loop, with the statement:

```
          X = AMAX1 (0.,(POTM(I)-TRSPOT))                    (5.19)
```

This defines X as the positive difference between the matric potential of soil moisture at each layer and a threshold potential for root growth, which was assumed to be -10 bars in our simulation.

We now compute the daily birth and extension rates for the roots in each layer as a function of matric potential:

```
          BIRTH = BR*(1.-EXP(-AA*X**BB))                     (5.20)

          EXTENS = ER*(1.-EXP(-AA*X**BB))                    (5.21)
```

wherein BR and ER are the daily birth rate and extension rate coefficients, respectively; and AA, BB are constants of the sigmoid-shaped exponential dependence of both the birth and the extension rates upon the soil's matric potential. These rates start from zero at a matric potential equal to the threshold potential TRSPOT and approach the values of BR and ER, respectively, at a matric potential of zero (*i.e.*, as the soil approaches saturation). Aeration restrictions are disregarded in this context.

Next we determine the actual proliferation rate, after first setting it equal to zero:

```
          PROLIF = 0.
          IF(AVPOT(I).GE.TRSPOT)...
            PROLIF = BIRTH
```

Finally, we determine the length of roots per unit volume of soil (PRTL) as follows:

```
          PRTL(I) = PRTL(I)*(1.-DR)*(1.+PROLIF)                (5.22)

          IF(I.EQ.1)GO TO 1009
          IF(AVPOT(I-1).GE.TRSPOT.AND.AVPOT(I).GE.TRSPOT)...

          PRTL(I) = PRTL(I)+PRTL(I-1)*EXTENS                   (5.23)
```

The last statement accounts for the contribution of root extension to the total quantity of roots in each layer.

The following statements terminate the procedure and set AVPOT(I) equal to zero for subsequent runs:

```
     1009 CONTINUE
     1010 CONTINUE
          DO 1015 I = 1,NJ
     1015 AVPOT(I) = 0.
     3000 CONTINUE
```

Results and Discussion

The results of a series of simulation runs for eight different root growth patterns are shown in Figures 5.18 and 5.19. The depth distributions of these root systems at the end of a 10-day period of extraction from the unreplenished moisture reserve of a medium-textured uniform soil profile with an initial volumetric wetness of 25 percent are shown in Figure 5.18. Note that all root systems shown began with an identical depth distribution represented by curve 2, which also represents the stationary root system analyzed in our preceding section. Curve 1 represents a similarly nongrowing root system, but with a steady death rate of 2.5 percent per day. Curve 3 portrays a root system in which both proliferation and death take place, and curve 4 the hypothetical case of proliferation without death. None of the foregoing root systems were allowed to extend deeper into the soil profile.

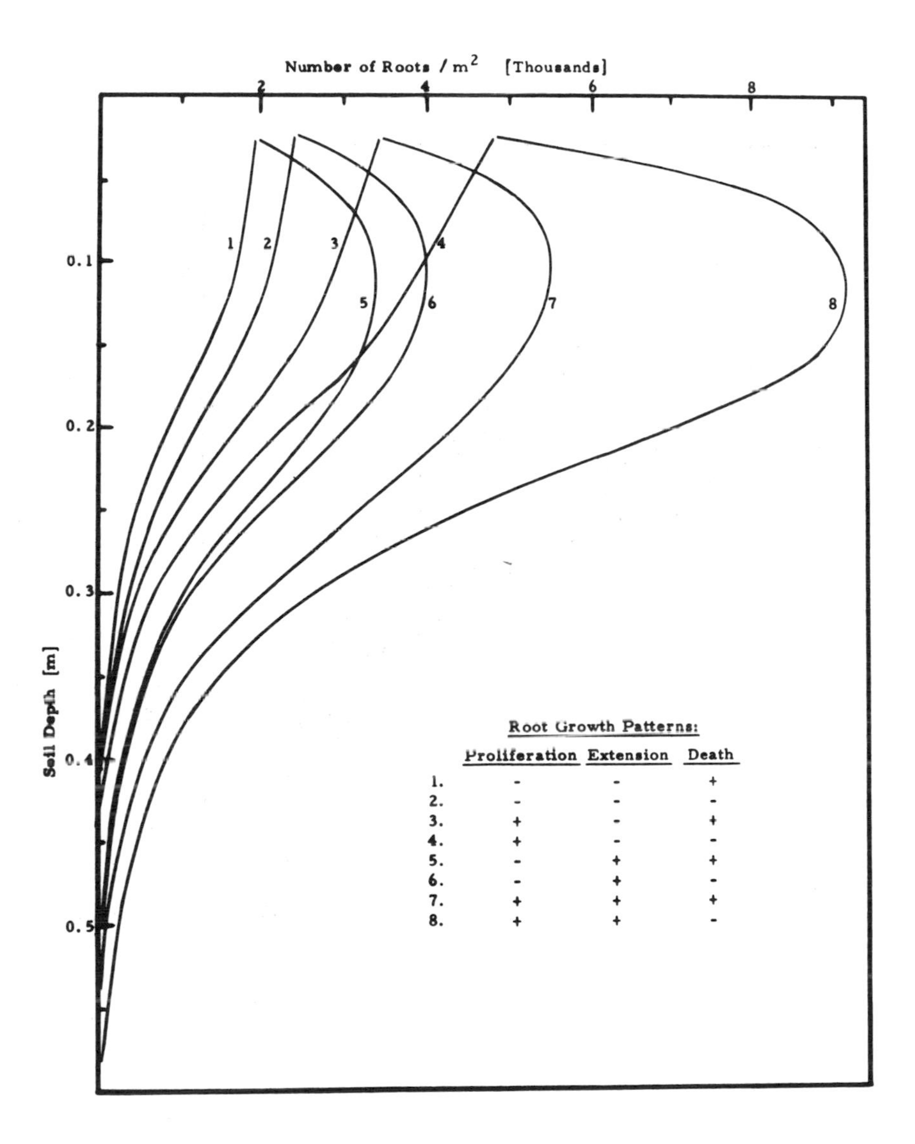

Figure 5.18. Depth distributions of initially uniform root systems at the end of a 10-day simulation of soil moisture extraction for eight different root growth patterns.

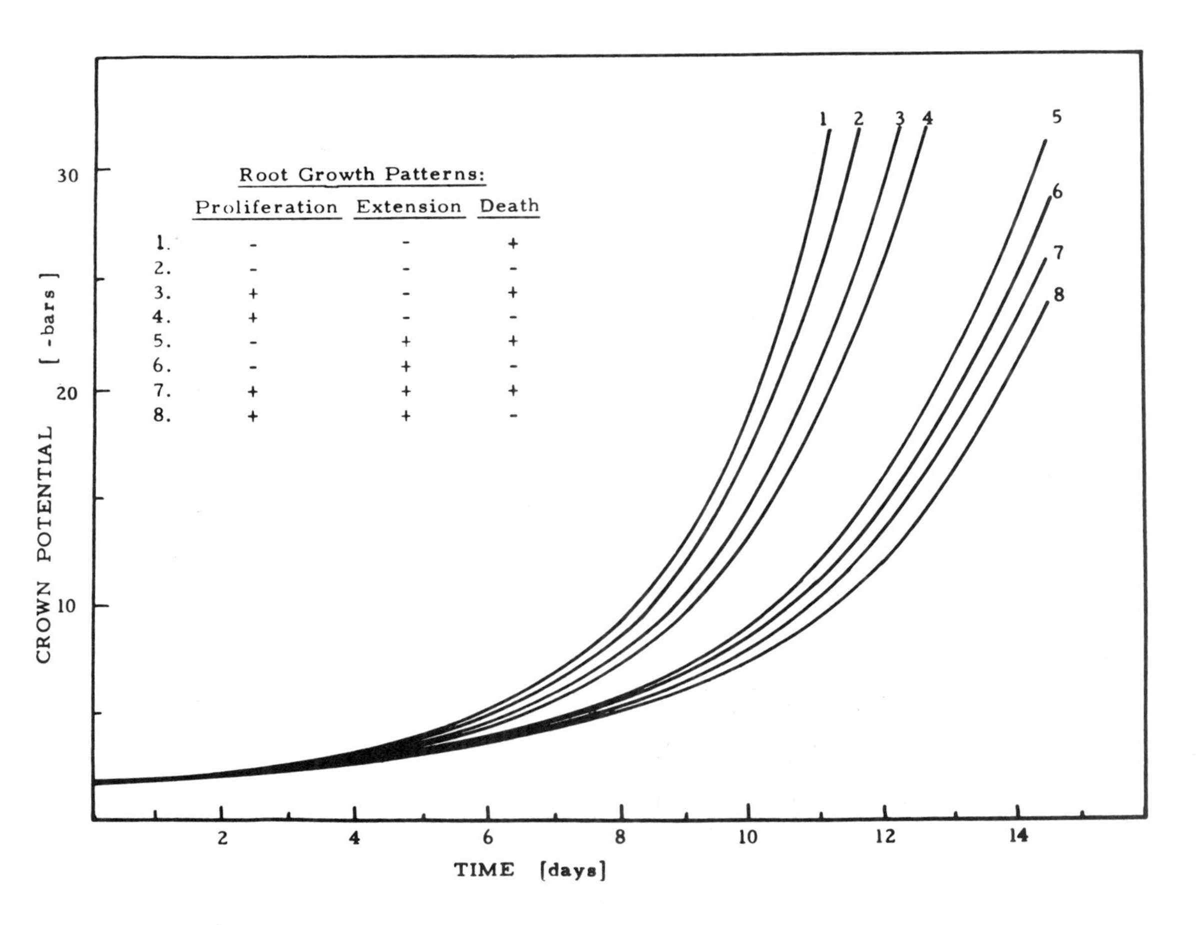

Figure 5.19. The time-course of plant water potential at the root crown during a 10-day period of soil moisture extraction by initially uniform root systems with eight different root growth patterns.

Curves 5, 6, 7, and 8 of Figure 5.18 portray the 10th-day distributions of corresponding root systems which differ from curves 1, 2, 3, 4, respectively, in the one important respect that they were allowed to extend into the soil profile, from layer to layer, according to Equations (5.21) and (5.23) at a variable rate which depended on the daily average value of matric potential. The difference between the two groups (curves 1, 2, 3, 4, versus 5, 6, 7, 8) is very significant. The extending root systems develop a distribution such that the maximum concentration of roots, initially near the soil surface, moves gradually down into the soil profile. As the upper layers are depleted of their moisture, by evaporation and root extraction, roots tend to grow preferentially into the profile so that the "center of gravity" of the extraction process moves progressively downward and thus the effective reserve of soil moisture available to the root system increases accordingly (see Figure 5.20).

The possible effect of root growth processes on plant water relations is indicated in Figure 5.19, which shows the time-course of plant water potential at the "crown" of the roots (where all the roots converge and the stem emerges from the soil with a single value of water potential). Like its predecessor, Figure 5.19 shows a distinct separation between the group of nonextending root systems capable of penetrating into progressively deeper layers in the soil profile (represented by curves 5, 6, 7, 8). Without extension, root proliferation added only about 1 day to the period of time the plant could maintain the potential transpiration rate without experiencing excessive stress (*i.e.*, a crown potential lower than -30 bars). On the other hand, the root systems capable of extending themselves deeper into the soil profile prolonged that time span by at least 3 days.

The phenomenon of root extension is characteristic of a stand of young plants, such as an annual crop in its early stages of growth. The rate of root system extension probably decreases in older plants and may indeed become negligible in the case of mature perennial crops such as alfalfa or Rhodes grass for which the root zone depth eventually reaches a limit. Our model makes no provision for this time-dependent aspect of root extension, nor does it account for the possible restriction of soil aeration which may play an important role in limiting root penetration into the deeper soil layers. In principle, however, such factors can be incorporated into this sort of model without undue difficulty, given the quantitative relationships involved and the parameters to characterize them. On the other hand, the very ease with which theoretical models can be developed into increasingly complex hypothetical constructions without any apparent logical limit presents a problem in itself. The imagination of modelers and the capability of computers already exceed the bounds of our experimental information on the behaviour of the real system which we may pretend to simulate. However much we believe our own

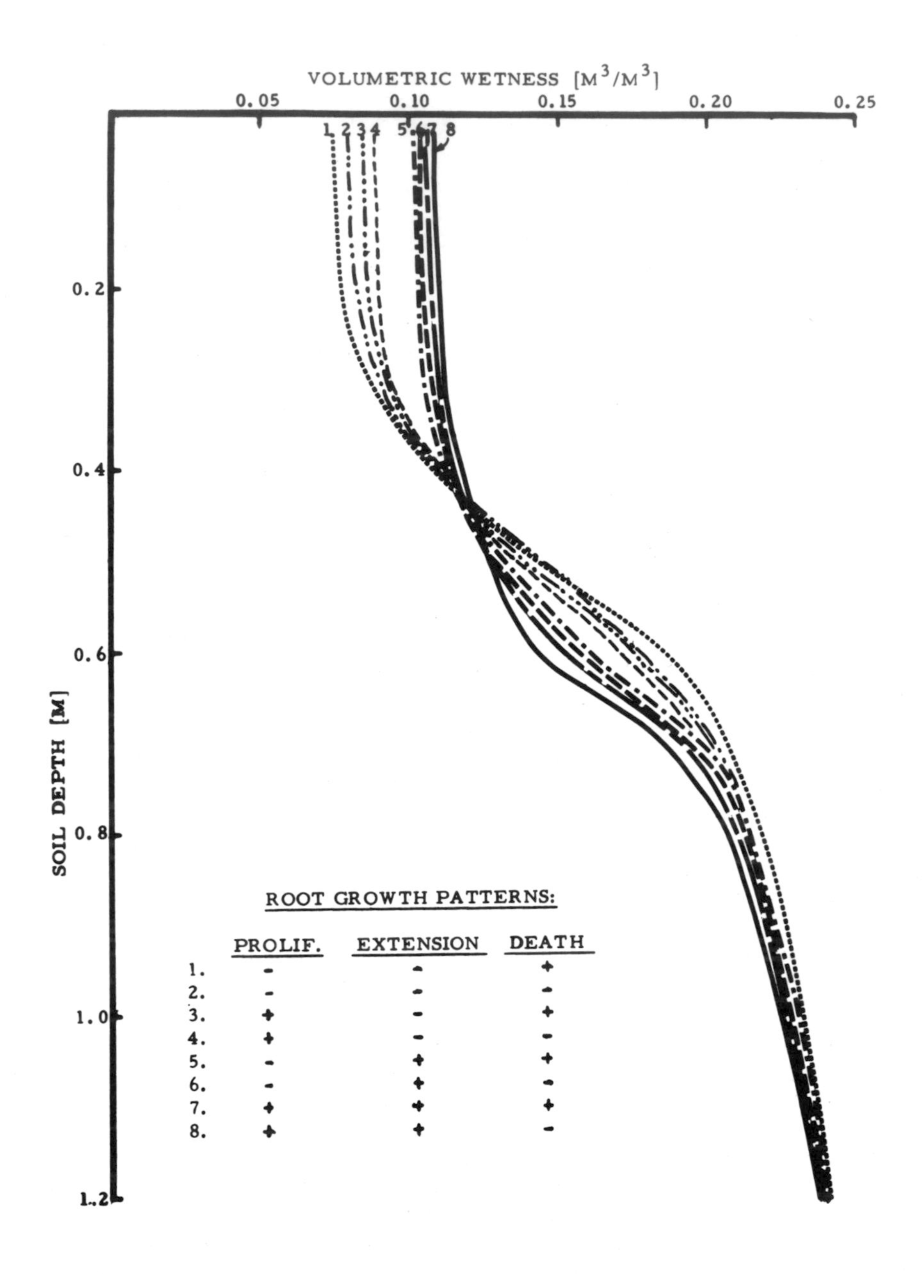

Figure 5.20. Soil moisture profiles at the end of a 10-day simulation of moisture extraction by root systems with various growth patterns.

model to be based on essentially sound concepts of soil moisture and root system dynamics, it still requires rigorous testing, which is indeed a very arduous and painstaking task. The sort of theoretical thought-experiment described herein will only serve its purpose if it helps to spur the task of experimentation and the analysis of real data.

EPILOGUE

The task of simulation in the area of environmental physics requires the combined efforts of specialists in numerical mathematics and systems analysis and of specialists in the natural and experimental sciences of soil physics, micrometeorology, plant physiology, and hydrology. Experts in the techniques of computer modeling may not always know just where their expertise can and ought to be applied to best advantage. Environmental scientists know the problems in the field, but may not be fully aware of the possibilities inherent in the simulation approach toward the solution of problems which have thus far defied traditional methods of analysis. The two are, on the one hand, like a professional actor, always in search of a good play in which to perform; and an undiscovered playwright, yearning for a good performer to redeem his plot. It is only when, by good fortune, the twain meet and join forces that we can expect a really good show.

The author of this monograph is a soil physicist, who, unlike his younger colleagues, received his basic training in the primitive prehistoric dark ages before computers appeared on the scene and became so ubiquitous and inescapable. Hence he has had to make an effort to re-train himself in the new art (or suffer the consequences of prematurely becoming an administrator!). The fact that he has been at least partially successful should be encouraging to others in the same predicament. In any case, the examples given should therefore not be taken as definitive solutions but only as demonstrations of some of the potentialities inherent in the simulation approach as applicable to soil physics.

The traditional tendency had been to isolate phenomena and study them separately in arbitrarily disjointed segments of the field environment. For instance, many agronomists, in their single-minded pursuit of greater crop yields (in itself a very worthy objective) had been myopically unaware that some of their fertilizers and pesticides may contaminate the environment outside their particular field. The timely appearance of computers and the simulation approach offer a way to re-integrate our fragmentary knowledge and thus overcome self-imposed artificial boundaries between heretofore separate disciplines, the adherents of which had become strangers to one another through the development of separate jargons and habits of thought.

However, the simulation approach has its pitfalls too. Computer simulation has become almost too easy, and there is danger of wanton use. Like any other tool, it can be abused (a knife, for instance, can be used to save a life — or do the opposite). A modeler can become so enamoured of his tool, its elegance, its ease of manipulation, that he may become addicted to it. After a beautiful exercise in simulation, producing such neat and definitive results, the tedious and painstaking process of experimentation may well seem like an anticlimax. Even worse, the modeler may develop such a vested interest in the success of his creation, that he might end up, like Dr. Frankenstein, its victim (at least in the sense of losing his objectivity).

It must be remembered that simulation *per se* cannot solve a real problem. It can only *simulate* a solution. Its results are predetermined by the input, although the full consequences of this determinism are often unforeseen for complex systems. A simulation can indeed provide new perspectives on the problem, but its predictions are always somewhat doubtful, even when the model is basically sound. When a model is not sound, *i.e.,* when its premises or data are wrong, there is great danger that it will gain a false aura of respectability merely because it was processed on a computer, which still conveys a sense of magic to many laymen. Simulation is not a panacea. It is not a substitute for experimentation, but a possibly more rational basis for experimentation. We need detailed, sound, and comprehensive experimentation as a basis for devising models, as well as for supplying the necessary parameters, and for validating (or refuting!) their results. Reciprocally, such results can help economize experimentation by guiding it to where it is needed most.

Because of the limited accuracy of all measurements, it is often possible that several models differing in structure may give equally good predictions within experimental errors. If only compatibility to data were of interest, any of the models would serve equally well, and perhaps statistical models are preferable as they are based on the generally safe premise that the future will probably resemble the past. But such models do not enlighten and lead nowhere. If we wish to understand how the system works, then we should prefer the models which are based on the fundamental mechanisms known to operate within the system. The scientific emphasis perhaps ought to be not on the question of a model's uniqueness or even on how closely its predictions fit a particular set of data, but on the clarity with which the discrepancies between model predictions and experimental data might lead to new and profitable inferences. There is no such thing as *the* model, capable of incorporating *all* the system's complexities, only *a* model, one at a time, each with a limited purpose. After all, we do not seek to assemble and wrap up all knowledge in a final sealed package, but rather to discover the missing facts.

Simulation can be deceptive. It is always tentative. It is never exact. It can hardly generate any new information. We must not expect too much of it. Its promise so far has exceeded its performance. It is still very much a moot question as to whether the art of simulation is already ripe to serve as a guide in the actual practice of soil and water management. Yet, even in its present state, simulation seems to offer a good point of departure for theorists and experimentalists to begin journeying together.

"Therefore theory, which gives facts their value and significance, is often very useful, even if it is partially false, because it throws light on phenomena which no one has observed, it forces an examination, from many angles, of facts which no one has hitherto studied, and provides the impulse for more extensive and more productive researches....

It is a moral duty for the man of science to expose himself to the risk of committing error, and to submit to criticism in order that science may continue to progress.... Those who are endowed with a mind serious and impersonal enough not to believe that everything they write is the expression of absolute and eternal truth will approve of this theory, which puts the aims of science well above the miserable vanity and paltry *amour propre* of the scientist."

G. Ferrero, *Les Lois psychologiques du symbolisme*, Paris, 1895.

"Science is the art of creating suitable illusions which the fool believes or argues against, but the wise man enjoys for their beauty or their ingenuity, without being blind to the fact that they are human veils and curtains concealing the abysmal darkness of the Unknowable."

C.G. Jung

REFERENCES

Amerman, C.R. (1973). Hydrology and soil science. In: "Field Soil Water Regime." (Bruce, R.R., *et al.*, Editors). Special Publication No. 5, Soil Sci. Soc. Amer., Madison, Wisconsin.

Arnold, G.W. and de Wit, C.T., Editors (1976). "Critical Evaluation of Systems Analysis in Ecosystems Research and Management." Centre for Agricultural Publishing and Documentation (PUDOC). Wageningen, Netherlands.

Aylor, D.E., and Parlange, J-Y. (1973). Vertical infiltration into a layered soil. Soil Sci. Soc. Amer. Proc. 37: 673-676.

Baver, L.D. (1940). "Soil Physics." John Wiley & Sons, New York,

Beek, J. and Frissel, M.J. (1973). "Simulation of Nitrogen Behavior in Soils." Simulation Monographs. Centre for Agricultural Publishing and Documentation (PUDOC). Wageningen, Netherlands.

Black, Max (1962). "Models and Metaphors." Cornell Univ. Press, Ithaca, New York.

Black, T.A., Gardner, W.R., and Thurtell, G.W. (1969). The prediction of evaporation, drainage and soil water storage for bare soil. Soil Sci. Soc. Amer. Proc. 33: 655-660.

Boast, C.W. (1973). Modeling the movement of chemicals in soils by water. Soil Science. 115: 224-230.

Brennan, R.D. (1968). Continuous system modeling programs. State-of-the-art and prospects for development. Proceedings for the IFIP Working Conference on Simulation Languages. Oslo, 1967. North-Holland Publishing Co., Amsterdam.

Brennan, R.D. and Silberberg, M.Y. (1968). The System 360 continuous system modeling program. Simulation 11: 301-308.

Bresler, E. (1973). Simultaneous transport of solutes and water under transient unsaturated flow conditions. Water Resources Research 9: 975-986.

Bresler, E., Kemper, W.D., and Hanks, R.J. (1969). Infiltration, redistribution, and subsequent evaporation of water from soil as affected by wetting rate and hysteresis. Soil Sci. Soc. Amer. Proc. 33: 832-840.

Brouwer, R. (1965). Water movement across the root. Symp. Soc. Exp. Biol. 19: 131-149.

Buxton, J.N., Editor. (1968). "Simulation Programming Languages." Proceedings IFIP Working Conference, Oslo, 1967. North-Holland Publishing Company, Amsterdam.

Childs, E.C. and Collis-George, N. (1950). The permeability of porous materials. Proc. Roy. Soc. 201A: 392-405.

Cochran, W.G. and Cox, G.M. (1957). "Experimental Design." John Wiley & Sons, New York.

Cohen, K.J. and Cyert, R.M. (1961). Computer models in dynamic economics. The Quarterly Journal of Economics. LXXV: 112-127.

Cowan, I.R. (1965). Transport of water in the soil-plant-atmosphere system. J. Appl. Ecol. 2: 221-239.

Cowan, I.R. and Milthorpe, F.L. (1968). Plant factors influencing the water status of plant tissues. In: Water Deficits and Plant Growth (Kozlowski, T.T., Editor). Academic Press, New York.

Davidson, J.M., Nielsen, D.R. and Biggar, J.W. (1966). The dependence of soil water uptake and release upon the applied pressure increment. Soil Sci. Soc. Amer. Proc. 30: 298-304.

de Vries, D.A. (1966). Thermal properties of soils. In: "Physics of Plant Environment." (van Wijk, W.R., Editor). North-Holland Publishing Co., Amsterdam.

de Vries, D.A. (1975). Heat transfer in soils. In: "Heat and Mass Transfer in the Biosphere." (de Vries, D.A. and Afgan, N.H. Editors). John Wiley & Sons, New York.

de Vries, D.A. and Afgan, N.H. (1975). "Heat and Mass Transfer in the Biosphere." Scripta Book Co., John Wiley & Sons, New York.

de Wit, C.T. and van Keulen, H. (1972). "Simulation of Transport Processes in Soils." Centre for Agricultural Publishing and Documentation (PUDOC). Wageningen, Netherlands.

de Wit, C.T. and Goudriaan, J. (1974). "Simulation of Ecological Processes." Centre for Agricultural Publishing and Documentation (PUDOC). Wageningen, Netherlands.

Emerson, W.W. (1954). Water conduction by several grass roots. J. Agri. Sci. 45: 241-245.

Engman, E.T. (1974). Partial area hydrology and its application to water resources. Water Resources Bulletin. 10: 512-521.

Engman, E.T. and Rogowski, A.S. (1972). A partial area model for storm flow synthesis. Water Resources Research. 10: 464-472.

Feodoroff, A. and Rafi, M. (1962). Evaporation of water from bare soil. C.R. Acad. Sci. Paris. 255: 3220-3222.

Fisher, R.A. (1944). "Statistical Methods for Research Workers." Oliver & Boyd, London.

Forchheimer, P. (1930). "Hydraulik," Teubner, Leipzig and Berlin.

Freeze, R.A. (1972). Role of subsurface flow in generating surface runoff. 2. Upstream source areas. Water Resources Research. 8: 1272-1283.

Gardner, W.R. (1959). Solutions of the flow equation for the drying of soils and other porous media. Soil Sci. Soc. Amer. Proc. 23: 183-187.

Gardner, W.R. (1960). Dynamic aspects of water availability to plants. Soil Sci. 89: 63-67.

Gardner, W.R. (1964). Relation of root distribution to water uptake and availability. Agron. J. 56: 41-45.

Gardner, H.R. (1973). Prediction of evaporation from homogeneous soil based on the flow equation. Soil Sci. Soc. Amer. Proc. 37: 513-516.

Gardner, W.R. and Hillel, D.I. (1962). The relation of external evaporative conditions to the drying of soils. J. Geophys. Res. 67: 4319-4325.

Haines, W.B. (1930). Studies in the physical properties of soils: V. The hysteresis effect in capillary properties and the modes of moisture distribution associated therewith. Jour. Agr. Sci. 20: 97-116.

Hanks, R.J. and Bowers, S.B. (1962). Numerical solution of the moisture flow equation for infiltration into layered soils. Soil Sci. Soc. Amer. Proc. 26: 530-534.

Hanks, R.J. and Gardner, H.R. (1965). Influence of different diffusivity water content relations on evaporation of water from soils. Soil Sci. Soc. Amer. Proc. 495-498.

Hide, J.C. (1954). Observations on factors influencing the evaporation of soil moisture. Soil Sci. Soc. Amer. Proc. 18: 234-239.

Hillel, D. (1971). "Soil and Water: Physical Principles and Processes." Academic Press, New York.

Hillel, D. (1974). Methods of laboratory and field investigation of physical properties of soils. Trans. 10th Int. Congr. Soil Sci. I: 301-308. Moscow.

Hillel, D. (1975). Some criteria for comprehensive modeling of transport phenomena in the soil-plant-atmosphere continuum. In: "Computer Simulation of Water Resources Systems." (Vansteenkiste, G.C., Editor). North-Holland Publishing Co., Amsterdam.

Hillel, D. (1975). Evaporation from bare soil under steady and diurnally fluctuating evaporativity. Soil Sci. 120: 230-237.

Hillel, D. (1976). On the role of soil moisture hysteresis in the suppression of evaporation from bare soil. Soil Sci. 122: 309-314.

Hillel, D. and Tadmor, N.H. (1962). Water regime and vegetation of principal plant habitats in the Negev Highlands of Israel. Ecology 43: 33-41.

Hillel, D. and Gardner, W.R. (1969). Steady infiltration into crust-topped profiles. Soil Sci. 108: 137-142.

Hillel, D., Krentos, V., and Stylianou, Y. (1972). Procedure and test of an internal drainage method for measuring soil hydraulic characteristics *in situ*. Soil Sci. 114: 395-400.

Hillel, D. and Benjamini, Y. (1974). Experimental comparison of infiltration and drainage methods for determining unsaturated hydraulic conductivity of a soil profile *in situ*. Proc. FAO/IAEA Symposium on Isotopes and Radiation Techniques in Studies of Soil Physics. Vienna.

Hillel, D. and Berliner, P. (1974). Waterproofing surface-zone soil aggregates for water conservation. Soil Sci. 118: 131-135.

Hillel, D., van Bavel, C.H.M., and Talpaz, H. (1975a). Dynamic simulation of water storage in fallow soil as affected by mulch of hydrophobic aggregates. Soil Sci. Soc. Amer. Proc. 39: 826-833.

Hillel, D., van Beek, C., and Talpaz, H. (1975b). A microscopic model of soil water uptake and salt movement to plant roots. Soil Sci. 120: 385-399.

Hillel, D., Talpaz, H., and van Keulen, H. (1975c). A macroscopic scale model of water uptake by a non-uniform root system and of water and salt movement in the soil profile. Soil Sci. 121: 242-255.

Hillel, D. and Talpaz, H. (1976). Effect of root growth parameters on the pattern of soil moisture extraction by non-uniform root systems. Soil Sci. 121: 307-312.

Hillel, D. and van Bavel, C.H.M. (1976). Dependence of profile water storage on soil texture and hydraulic properties: a simulation model. Soil Sci. Soc. Amer. J. 40: 807-815.

Hillel, D. and Talpaz, H. (1977). Simulation of soil water dynamics in layered soils. Soil Sci. 123:54-62.

Hyndman, D.E. (1970). "Analog and Hybrid Computing." Pergamon Press, Oxford.

IBM Corporation. (1972). System/360 Continuous Systems Modeling Program. User's Manual, 5th edition, GH20-0367-4. Data Processing Division, IBM, White Plains, New York 10604.

Idso, S.B., Reginato, R.J., Jackson, R.D., Kimball, B.A., and Nakayama, F.S. (1974). The three stages of drying a field soil. Soil Sci. Soc. Amer. Proc. 38: 831-837.

Jackson, R.A. (1972). On the calculation of hydraulic conductivity. Soil Sci. Soc. Amer. Proc. 36: 380-383.

Jackson, R.D. (1973). Diurnal changes in soil water content during drying. In: "Field Soil Water Regime." Soil Sci. Soc. Amer., Madison, Wisconsin, pp. 37-55.

Jackson, R.D., Kimball, B.A., Reginato, R.J., and Nakayama, S.F. (1973). Diurnal soil water evaporation: Time-depth-flux patterns. Soil Sci. Soc. Amer. Proc. 37: 505-509.

Jackson, R.D., Reginato, R.J., Kimball, B.A., and Nakayama, F.S. (1974). Diurnal soil-water evaporation: comparison of measured and calculated soil-water fluxes. Soil Sci. Soc. Amer. Proc. 38: 861-866.

Keen, B.A. (1931). "The Physical Properties of the Soil." John Wiley & Sons, New York.

Kunze, R.J., Uchara, G., and Graham, K. (1968). Factors important in the calculation of hydraulic conductivity. Soil Sci. Soc. Amer. Proc. 32: 760-765.

Lambert, J.R. and Penning de Vries, F.W.T. (1973). Dynamics of water in the soil-plant-atmosphere system: A model named troika.

In: "Physical Aspects of Soil Water and Salts in Ecosystems." Springer-Verlag, Berlin.

Lemon, E.R. (1956). The potentialities for decreasing soil moisture evaporation loss. Soil Sci. Soc. Amer. Proc. 20: 120-125.

Leyser, J.P. and Loch, J.P.G. (1972). Effect of xylem resistance on the water relations of plant and soil. Department of Theoretical Production Ecology. Agricultural University, Wageningen, Netherlands.

Makkink, G.F. and van Heemst, H.D.J. (1975). "Simulation of the Water Balance of Arable Land and Pastures." Centre for Agricultural Publishing and Documentation (PUDOC). Wageningen, Netherlands.

Manning, R. (1891). On the flow of water in open channels and pipes. Trans. Inst. Civil Engng. Ireland 20: 161-207.

Marshall, T.J. (1958). A relation between permeability and size distribution of pores. J. Soil Sci. 9: 1-8.

Martin, F.F. (1968). "Computer Modeling and Simulation." John Wiley & Sons, New York.

Miller, E.E. and Miller, R.D. (1956). Physical theory for capillary flow phenomena. J. Appl. Phys. 27: 324-332.

Millington, R.J. and Quirk, J.P. (1959). Permeability of porous media. Nature 183: 387-388.

Molz, F.J., Remson, I., Fungaroli, A.A., and Drake, R.L. (1968). Soil moisture availability for transpiration. Water Resources Research 4: 1161-1169.

Molz, F.J. and Remson, I. (1970). Extraction-term models of soil moisture use by transpiring plants. Water Resources Research 6: 1346-1356.

Molz, F.J. and Remson, I. (1971). Application of an extraction-term model to the study of moisture flow to plant roots. Agron. J. 63: 72-77.

Monteith, J.L. (1963). Gas exchange in plant communities. In: "Environmental Control of Plant Growth." (Evans, L.T., Editor). Academic Press, New York.

Muskat, M. (1946). "The Flow of Homogeneous Fluids Through Porous Media." Edwards, Ann Arbor, Michigan.

Naylor, T.H., Balintfy, J.L., Burdick, D.S., and Chu, K. (1968). "Computer Simulation Techniques." John Wiley & Sons, New York and London.

Newman, E.I. (1969). Resistance to water flow in soil and plant. J. Appl. Ecol. 15: 1-12.

Nielsen, D.R. and Biggar, J. (1962). Miscible displacement: III. Theoretical considerations. Soil Sci. Soc. Amer. Proc. 28: 216-221.

Nielson, D.R., Biggar, J.W., and Erh, K.T. (1973). Spatial variability of field-measured soil-water properties. Hilgardia 42: 215-259.

Nimah, A and Hanks, R.J. (1973). Model for estimating soil water, plant, and atmospheric interrelations: I. Description and sensitivity. Soil Sci. Soc. Amer. Proc. 37: 522-527.

Ogata, Gen., Richards, L.A., and Gardner, W.R. (1960). Transpiration of alfalfa determined from soil water content changes. Soil Sci. 89: 179-182.

Philip, J.R. (1957). The physical principles of water movement during the irrigation cycle. Proc. 3rd Int. Congr. Irrig. Drainage 8: 125-128, 154.

Philip, J.R. (1957). Evaporation, and moisture and heat fields in the soil. J. Meteorol. 14: 354-366.

Philip, J.R. (1966). Plant water relations: some physical aspects. Ann. Rev. Plant Physiol. 17: 245-268.

Philip, J.R. (1967). The second stage of drying of soil. J. Appl. Meteorol. 6: 581-582.

Philip, J.R. (1975). Stability analysis of infiltration. Soil Sci. Soc. Amer. Proc. 39: 1042-1049.

Philip, J.R. and de Vries, D.A. (1957). Moisture movement in porous materials under temperature gradients. Trans. Amer. Geophys. Union 38: 222-232.

Poulovassilis, A. (1962). Hysteresis of pore water, an application of the concept of independent domains. Soil Sci. 93: 405-412.

Raats, P.A.C. (1973). Unstable wetting fronts in uniform and nonuniform soils. Soil Sci. Soc. Amer. Proc. 37: 681-685.

Reichenbach, H. (1951). "The Rise of Scientific Philosophy." University of California Press, Berkeley.

Ripple, C.D, Rubin, J, and vah Hylkama, T.E.A. (1972). Estimating steady-state evaporation rates from bare soils under conditions of high water table. U.S. Geol. Sur., Water Supp. Pap. 2019-A.

Rubin, J. (1967). Numerical method for analyzing hysteresis-affected, post-infiltration redistribution of soil moisture. Soil Sci. Soc. Amer. Proc. 31: 13-20.

Sellers, W.D. (1965). "Physical Climatology." Univ. Chicago Press, Chicago.

Sellin, R.H.J. (1969). "Flow in Channels." MacMillan & Co., London.

Slatyer, R.O. (1967). "Plant-water Relationships." Academic Press, London.

Smith, R.E. and Woohiser, D.A. (1971). Mathematical simulation of infiltrating watershed. Hydrology Papers. Colorado State University, Fort Collins, Colorado. 44 pp.

Staple, W.J. (1969). Comparison of computed and measured moisture redistribution following infiltration. Soil Sci. Soc. Amer. Proc. 33: 840-847.

Staple, W.J. (1974). Modified Penman equation to provide the upper boundary conditions in computing evaporation from soil. Soil Sci. Soc. Amer. Proc. 38: 837-839.

Stroosnijder, L., van Keulen, H., and Vachaud, G. (1972). Water movement in layered soils: 2. Experimental confirmation of a simulation model. Neth. J. Agri. Sci. 20: 67-72.

Swartzendruber, D. and Hillel, D. (1975). Infiltration and runoff for small field plots under constant intensity rainfall. Water Resources Research 11: 445-451.

Szceiz, G., van Bavel, C.H.M., and Takami, S. (1973). Stomatal factor in the water use and dry matter production by sorghum. Agric. Meteorology 12: 361-389.

Takagi, S. (1960). Analysis of the vertical downward flow of water through a two-layered soil. Soil Sci. 90: 98-103.

Taylor, H.M., Huck, M.G., and Klepper, B. (1972). Root development in relation to soil physical conditions. In: "Optimizing the Soil Physical Environment Toward Greater Crop Yields." (Hillel, D., Editor). Academic Press, New York.

Taylor, H.M. and Klepper, B. (1975). Water uptake by cotton root systems: An examination of assumptions in the single root model. Soil Sci. 120: 57-67.

Vachaud, G. and Thony, J.L. (1971). Hysteresis during infiltration and redistribution in a soil column at different initial water contents. Water Resources Research 7: 111-127.

van Bavel, C.H.M. (1966). Potential evaporation: The combination concept and its experimental verification. Water Resources Research 2: 455-467.

van Bavel, C.H.M. and Hillel, D. (1975). A simulation study of soil heat and moisture as affected by a dry mulch. Proceedings 1975 Summer Computer Simulation Conference, San Francisco, California. Simulation Councils, Inc., La Jolla, California.

van Bavel, C.H.M. and Hillel, D. (1977). Calculating potential and actual evaporation from a bare soil surface by simulation of concurrent flow of water and heat. Agric. Meteorology (in press).

van den Honert, T.H. (1948). Water transport in plants as a catenary process. Discuss. Faraday Soc. 3: 146-153.

van der Ploeg, R.R. and Benecke, P. (1974). Unsteady, unsaturated, N-dimensional moisture flow in soil: A computer simulation program. Soil Sci. Soc. Amer. Proc. 38: 881-888.

van Keulen, H. (1975). "Simulation of Water Use and Herbage Growth in Arid Regions." Centre for Agricultural Publishing and Documentation (PUDOC). Wageningen, Netherlands.

van Keulen, H. and van Beek, C.G.E.M. (1971). Water movement in layered soils: A simulation model. Neth. J. Agr. Sci. 19: 138-153.

van Keulen, H. and Hillel, D. (1974). A simulation study of the drying-front phenomenon. Soil Sci. 118: 270-273.

van Schilfgaarde, J. (1974). "Drainage for Agriculture." Agron. Monograph 17, Amer. Soc. Agron., Madison, Wisconsin.

Wadleigh, C.H. (1946). The integrated soil moisture stress upon a root system in a large container of saline soil. Soil Sci. 6: 225-238.

Whisler, F.D. and Klute, A. (1965). The numerical analysis of infiltration, considering hysteresis, into a vertical soil column under gravity. Soil Sci. Soc. Amer. Proc. 29: 489-494.

Whisler, F.D., Klute, A., and Millington, R.J. (1968). Analysis of Steady state evapotranspiration from a soil column. Soil Sci. Soc. Amer. Proc. 32: 167-174.

Whisler, F.D. and Klute, A. (1969). Analysis of infiltration into stratified soil columns. In: "Water in the Unsaturated Zone." (Wageningen, Netherlands) IASH-AIHS-UNESCO 1: 451-472.

Wiegand, C.L. and Taylor, S.A. (1961). Evaporative drying of porous media: Influential factors, physical phenomena, mathematical analyses. Utah State Univ., Agr. Exp. Sta. Special Rep. No. 15.

Wierenga, P.J. and de Wit, C.T. (1970). Simulation of heat transfer in soils. Soil Sci. Soc. Amer. Proc. 34: 845-847.

Wind, G.P. (1955). Flow of water through plant roots. Neth. J. Agri. Sci. 3: 259-264.

Youngs, E.G. (1960). The hysteresis effect in soil moisture studies. Trans. 7th Intern. Soil Sci. Congr. Madison. 1: 107-113.

Zeigler, B.P. (1976). "Theory of Modeling and Simulation." John Wiley & Sons, New York.

"To write simply is as difficult as to be good."

W. Somerset Maugham